On the Shoulders of Giants

Harmonies of the World
Book Five

On the Shoulders of Giants

Harmonies of the World
Book Five

by Johannes Kepler
Edited, with Commentary, by Stephen Hawking

RUNNING PRESS
PHILADELPHIA · LONDON

9   8   7   6   5   4   3   2   1

Digit on the right indicates the number of this printing

Library of Congress Control Number: 2004094110

ISBN 0-7624-2018-9

Author photo courtesy of Book Laboratory

Cover Design: Doogie Horner

Typography: AGaramond, Weiss, Goudy

Cover images: Portrait of Johannes Kepler: Hulton/Archive by Getty Images; (bottom) Ring Nebula STScI-1999-01, courtesy of NASA and the Hubble Heritage Team (STScI/AURA/NASA)

Text of *Harmonies of the World* courtesy of Annapolis: St. John's Bookstore, © 1939.

This book may be ordered by mail from the publisher. Please include $2.50 for postage and handling.

But try your bookstore first!

Running Press Book Publishers

125 South Twenty-second Street

Philadelphia, PA 19103-4399

Visit us on the web!

www.runningpress.com

Contents

A Note to the Reader vii

Introduction ix

Johannes Kepler (1571–1630) His Life and Work xi

HARMONIES OF THE WORLD
BOOK FIVE

Proem 2

1. Concerning the Five Regular Solid Figures 4

2. On the Kinship Between the Harmonic Ratios
   and the Five Regular Figures 5

3. A Summary of Astronomical Doctrine Necessary for Speculation
   into the Celestial Harmonies 8

4. In What Things Having to do with the Planetary Movements Have the
   Harmonic Consonances been Expressed by the Creator, and in
   What Way? 18

5. In the Ratios of the Planetary Movements which are Apparent as it were to
   Spectators at the Sun, have been Expressed the Pitches of the System, or
   Notes of the Musical Scale, and the Modes of Song (Genera Cantus), the
   Major and the Minor 30

6. In the Extreme Planetary Movements the Musical Modes or Tones Have
   Somehow Been Expressed 35

7. The Universal Consonances of All Six Planets, Like Common Four-Part
   Counterpoint, Can Exist 37

8. In the Celestial Harmonies Which Planet Sings Soprano, Which Alto,
   Which Tenor, and Which Bass? 46

9. The Genesis of the Eccentricities in the Single Planets from the
   Procurement of the Consonances between their Movements 47

10. Epilogue Concerning the Sun, by way of Conjecture 83

Acknowledgments 91

About Stephen Hawking 92

A Note to the Reader

Harmonies of the World is one volume in a series of five scientific classics that Running Press has chosen to publish in a new format with commentary by renowned physicist, Stephen Hawking. Hawking chose these five particular essays to illuminate the evolution of modern physics and astronomy, as well as the process through which scientific knowledge evolves.

Also included in this series are:

- *On the Revolutions of Heavenly Spheres* by Nicolaus Copernicus
- *Dialogues Concerning Two New Sciences* by Galileo Galilei
- *Principia* by Sir Isaac Newton
- *Selections from* The Principle of Relativity by Albert Einstein

Look for all five essays combined into one volume in:

- *On the Shoulders of Giants* edited, with commentary, by Stephen Hawking
- *The Illustrated On the Shoulders of Giants* edited, with commentary, by Stephen Hawking

The text for *Harmonies of the World* is based on a translation of the original, printed edition. We have made no attempt to modernize the author's own distinct usage, spelling, or punctuation. The details of this essay are as follows:

We have selected Book Five of the five-book *Harmonies of the World* by Johannes Kepler. Kepler completed the work on May 27, 1618, publishing it under the title *Harmonies Mundi*. This translation is by Andrew Motte.

The Editors

Introduction

If I have seen farther, it is by standing on the shoulders of giants, wrote Isaac Newton in a letter to Robert Hooke in 1676. Although Newton was referring to his discoveries in optics rather than his more important work on gravity and the laws of motion, it is an apt comment on how science, and indeed the whole of civilization, is a series of incremental advances, each building on what went before. This is the theme of this fascinating series, which uses the original texts to trace the evolution of our picture of the heavens from the revolutionary claim of Nicolaus Copernicus that the earth orbits the sun to the equally revolutionary proposal of Albert Einstein that space and time are curved and warped by mass and energy. It is a compelling story because both Copernicus and Einstein have brought about profound changes in what we see as our position in the order of things. Gone is our privileged place at the center of the universe, gone are eternity and certainty, and gone are Absolute Space and Time to be replaced by rubber sheets.

It is no wonder both theories encountered violent opposition: the Inquisition in the case of the Copernican theory and the Nazis in the case of Relativity. We now have a tendency to dismiss as primitive the earlier world picture of Aristotle and Ptolemy in which the Earth was at the center and the Sun went round it. However we should not be too scornful of their model, which was anything but simpleminded. It incorporated Aristotle's deduction that the Earth is a round ball rather than a flat plate and it was reasonably accurate in its main function, that of predicting the apparent positions of the heavenly bodies in the sky for astrological purposes. In fact, it was about as accurate as the heretical suggestion put forward in 1543 by Copernicus that the Earth and the planets moved in circular orbits around the Sun.

Galileo found Copernicus's proposal convincing not because it better fit the observations of planetary positions but because of its simplicity and elegance, in contrast to the complicated epicycles of the Ptolemaic model. In *Dialogues Concerning Two New Sciences*, Galileo's characters, Salviati and Sagredo, put forward persuasive arguments in support of Copernicus. Yet, it was still possible for his third character, Simplicio, to defend Aristotle and Ptolemy and to maintain that in reality the Earth was at rest and the Sun went round the Earth.

It was not until Kepler's work made the Sun-centered model more accurate and Newton gave it laws of motion that the Earth-centered picture finally lost all credibility. It was quite a shift in our view of the universe: If we are not at the center, is our existence of any importance? Why should God or the Laws of Nature care about what happens on the third rock from the Sun, which is where Copernicus has left us? Modern scientists have out-Copernicused Copernicus by seeking an account of the universe in which Man (in the old pre-politically correct sense) played no role. Although this approach has succeeded in finding objective impersonal laws that govern the universe, it has not (so far at least) explained why the universe is the way it is rather than being one of the many other possible universes that would also be consistent with the laws.

Some scientists would claim that this failure is only provisional, that when we find the ultimate unified theory, it will uniquely prescribe the state of the universe, the strength of gravity, the mass and charge of the electron and so on. However, many features of the universe (like the fact that we are on the third rock, rather than the second or fourth) seem arbitrary and accidental and not the predictions of a master equation. Many people (myself included) feel that the appearance of such a complex and structured universe from simple laws requires the invocation of something called the anthropic principle, which restores us to the central position we have been too modest to claim since the time of Copernicus. The anthropic principle is based on the self-evident fact that we wouldn't be asking questions about the nature of the universe if the universe hadn't contained stars, planets and stable chemical compounds, among other prerequisites of (intelligent?) life as we know it. If the ultimate theory made a unique prediction for the state of the universe and its contents, it would be a remarkable coincidence that this state was in the small subset that allow life.

However the work of the last thinker in this series, Albert Einstein, raises a new possibility. Einstein played an important role in the development of quantum theory which says that a system doesn't just have a single history as one might have thought. Rather it has every possible history with some probability. Einstein was also almost solely responsible for the general theory of relativity in which space and time are curved and become dynamic. This means that they are subject to quantum theory and that the universe itself has every possible shape and history. Most of these histories will be quite unsuitable for the development of life but a very few have all the conditions needed. It doesn't matter if these few have a very low probability relative to the others: the lifeless universes will have no one to observe them. It is sufficient that there is at least one history in which life develops, and we ourselves are evidence for that, though maybe not for intelligence. Newton said he was *"standing on the shoulders of giants."* But as this series illustrates so well, our understanding doesn't advance just by slow and steady building on previous work. Sometimes as with Copernicus and Einstein, we have to make the intellectual leap to a new world picture. Maybe Newton should have said, *"I used the shoulders of giants as a springboard."*

Johannes Kepler

(1571–1630)

HIS LIFE AND WORK

If an award were ever given to the person in history who was most dedicated to the pursuit of absolute precision, the German astronomer Johannes Kepler might well be the recipient. Kepler was so obsessed with measurements that he even calculated his own gestational period to the minute—224 days, 9 hours, 53 minutes. (He had been born prematurely.) So it is no surprise that he toiled over his astronomical research to such a degree that he ultimately produced the most exact astronomical tables of his time, leading to the eventual acceptance of the sun-centered (heliocentric) theory of the planetary system.

Like Copernicus, whose work inspired him, Kepler was a deeply religious man. He viewed his continual study of universal properties as a fulfillment of his Christian duty to understand the very universe that God created. But unlike Copernicus, Kepler's life was anything but quiet and lacking in contrast. Always short of money, Kepler often resorted to publishing astrological calendars and horoscopes, which, ironically, gained him some local notoriety when their predictions turned out to be quite accurate. Kepler also suffered the early deaths of several of his children, as well as the indignity of having to defend in court his eccentric mother, Katherine, who had a reputation for practicing witchcraft and was nearly burned at the stake.

Kepler entered into a series of complex relationships, most notably with Tycho Brahe, the great naked-eye astronomical observer. Brahe dedicated years of his life to recording and measuring celestial bodies, but he lacked the mathematical and analytical skills necessary to understand planetary motion. A man of wealth, Brahe hired Kepler to make sense of his observations of the orbit of Mars, which had perplexed astronomers for many years. Kepler painstakingly mapped Brahe's data on the motion of Mars to an ellipse, and this success lent mathematical credibility to the Copernican model of a sun-centered system. His discovery of elliptical orbits helped usher in a new era in astronomy. The motions of planets could now be predicted.

In spite of his achievements, Kepler never gained much wealth or prestige and was often forced to flee the countries where he sojourned because of religious upheaval and civil unrest. By the time he died at the age of fifty-nine in 1630 (while attempting to collect an overdue salary), Kepler had discovered three laws of planetary motion, which are still taught to students in physics classes in the twenty-first century. And it was Kepler's Third Law, not an apple, that led Isaac Newton to discover the law of gravitation.

Johannes Kepler was born on December 27, 1571, in the town of Weil der Stadt, in Wurttemburg (now part of Germany). His father, Heinrich Kepler, was, according to Johannes, "an immoral, rough, and quarrelsome soldier" who deserted his family on several occasions to join up with mercenaries to battle a Protestant uprising in Holland. Heinrich is believed to have died somewhere in the Netherlands. The young Johannes lived with his mother, Katherine, in his grandfather's inn, where he was put to work at an early age waiting tables, despite his poor health. Kepler had nearsightedness as well as double vision, which was believed to have been caused by a near-fatal bout of small-pox; and he also suffered from abdominal problems and "crippled" fingers that limited his career potential choice, in the view of his family, to a life in the ministry.

"Bad-tempered" and "garrulous" were words Kepler used to describe his mother, Katherine, but he was aware from a young age that his father was the cause. Katherine herself had been raised by an aunt who practiced witchcraft and was burned at the stake. So it was no surprise to Kepler when his own mother faced similar charges later in her life. In 1577, Katherine showed her son the "great comet" that appeared in the sky that year, and Kepler later acknowledged that this shared moment with his mother had a lasting impact on his life. Despite a childhood filled with pain and anxiety, Kepler was obviously gifted, and he managed to procure a scholarship reserved for promising male children of limited means who lived in the German province of Swabia. He attended the German Schreibschule in Leonberg before transferring to a Latin school, which was instrumental in providing him with the Latin writing style he later employed in his work. Being frail and precocious, Kepler was beaten regularly by classmates, who considered him a know-it-all, and he soon turned to religious study as a way of escaping his predicament.

In 1587, Kepler enrolled at Tubingen University, where he studied theology and philosophy. He also established himself there as a serious student of mathematics and astronomy, and became an advocate of the controversial Copernican heliocentric theory. So public was young Kepler in his defense of the Copernican model of the universe that it was not uncommon for him to engage in public debate on the subject. Despite his main interest in theology, he was growing more and more intrigued by the mystical appeal of a heliocentric universe. Although he had intended to graduate from Tubingen in 1591 and join the university's theology faculty, a recommendation to a post in mathematics and astronomy at the Protestant school in Graz, Austria, proved irresistible. So at the age of twenty-two, Kepler deserteda career in the ministry for the study of science. But he would never abandon his belief in God's role in the creation of the universe.

In the sixteenth century, the distinction between astronomy and astrology was fairly ambiguous. One of Kepler's duties as a mathematician in Graz was to compose an astrological calendar complete with predictions. This was a common practice at the time, and Kepler was clearly motivated by the extra money the job provided but he could not have anticipated the public's reaction when his first calendar was published. He predicted an extraordinarily cold winter, as well as a Turkish incursion, and when both predictions came true, Kepler was triumphantly hailed as a prophet. Despite the clamor, he would never hold much respect for the work he did on the annual almanacs. He called astrology "the foolish little daughter of astronomy" and was equally dismissive of the public's interest and the astrologer's intentions. "If ever astrologers are correct," he wrote, "it ought to be credited to luck." Still, Kepler never failed to turn to astrology whenever money became tight, which was a recurring theme in his life, and he did hold out hope of discovering some true science in astrology.

One day, while lecturing on geometry in Graz, Kepler experienced a sudden revelation that set him on a passionate journey and changed the course of his life. It was, he felt, the secret key to understanding the universe. On the blackboard, in front of the class, he drew an equilateral triangle within a circle, and another circle drawn within the triangle. It occurred to him that the ratio of the circles was indicative of the ratio of the orbits of Saturn and Jupiter. Inspired by this revelation, he assumed that all six planets known at the time were arranged around the sun in such a way that the geometric figures would fit perfectly between them. Initially he tested this hypothesis without success, using two-dimensional plane figures such as the pentagon, the square, and the triangle. He then returned to the Pythagorean solids, used by the ancient Greeks, who discovered that only five solids could be constructed from regular geometric figures. To Kepler, this explained why there could only be six planets (Mercury, Venus, Earth, Mars, Jupiter, and Saturn) with five spaces between them, and why these spaces were not uniform. This geometric theory regarding planetary orbits and distances inspired Kepler to write *Mystery of the Cosmos* (*Mysterium Cosmographicum*), published in 1596. It took him about a year to write, and although the scheme was reasonably accurate, he was clearly very sure that his theories would ultimately bear out:

> *And how intense was my pleasure from this discovery can never be expressed in words. I no longer regretted the time wasted. Day and night I was consumed by the computing, to see whether this idea would agree with the Copernican orbits, or if my joy would be carried away by the wind. Within a few days everything worked, and I watched as one body after another fit precisely into its place among the planets.*

Kepler spent the rest of his life trying to obtain the mathematical proof and scientific observations that would justify his theories. *Mystery of the Cosmos* was the first decidedly Copernican work published since Copernicus's own *On Revolutions*, and as a theologian and astronomer Kepler was determined to understand how and why God designed the universe. Advocating a heliocentric system had serious religious implications, but Kepler maintained that the sun's centrality was vital to God's design, as it kept the planets aligned and in motion. In this sense, Kepler broke with Copernicus's heliostatic system of a sun "near" the center and placed the sun directly in the center of the system.

Today, Kepler's polyhedrons appear impracticable. But although the premise of *Mystery of the Cosmos* was erroneous, Kepler's conclusions were still astonishingly accurate and decisive, and were essential in shaping the course of modern science. When the book was published, Kepler sent a copy to Galileo, urging him to "believe and step forth," but the Italian astronomer rejected the work because of its apparent speculations. Tycho Brahe, on the other hand, was immediately intrigued. He viewed Kepler's work as new and exciting, and he wrote a detailed critique in the book's support Reaction to *Mystery of the Cosmos*, Kepler would later write, changed the direction of his entire life.

In 1597, another event would change Kepler's life, as he fell in love with Barbara Müller, the first daughter of a wealthy mill owner. They married on April 27 of that year, under an unfavorable constellation, as Kepler would later note in his diary. Once again, his prophetic nature emerged as the relationship and the marriage dissolved. Their first two children died very young, and Kepler became distraught. He immersed himself in his work to distract himself from the pain, but his wife did not understand his pursuits. "Fat, confused, and simpleminded" was how he described her in his diary, though the marriage did last fourteen years, until her death in 1611 from typhus.

In September 1598, Kepler and other Lutherans in Graz were ordered to leave town by the Catholic archduke, who was bent on removing the Lutheran religion from Austria. After a visit to Tycho Brahe's Benatky Castle in Prague, Kepler was invited by the wealthy Danish astronomer to stay there and work on his research. Kepler was somewhat wary of Brahe, even before having met him. "My opinion of Tycho is this: he is superlatively rich, but he knows not how to make proper use of it, as is the case with most rich people," he wrote. "Therefore, one must try to wrest his riches from him."

If his relationship with his wife lacked complexity, Kepler more than made up for it when he entered into a working arrangement with the aristocratic Brahe. At first, Brahe treated the young Kepler as an assistant, carefully doling out assignments without giving him much access to detailed observational data. Kepler badly wanted to be regarded as an equal and given some independence, but the secretive Brahe wanted to use Kepler to establish his own model of the solar system—a non-Copernican model that Kepler did not support.

Kepler was immensely frustrated. Brahe had a wealth of observational data but lacked the mathematical tools to fully comprehend it. Finally, perhaps to pacify his restless assistant, Brahe assigned Kepler to study the orbit of Mars, which had confused the Danish astronomer for some time because it appeared to be the least circular. Kepler initially thought he could solve the problem in eight days, but the project turned out to take him eight years. Difficult as the research proved to be, it was not without its rewards, as the work led Kepler to discover that Mars's orbit precisely described an ellipse, as well as to formulate his first two "planetary laws," which he published in 1609 in *The New Astronomy*.

A year and a half into his working relationship with Brahe, the Danish astronomer became very ill at dinner and died a few days later of a bladder infection. Kepler took over the post of Imperial Mathematician and was now free to explore planetary theory without being constrained by the watchful eye of Tycho Brahe. Realizing an opportunity, Kepler immediately went after the Brahe data that he coveted before Brahe's heirs could take control of them. "I confess that when Tycho died," Kepler wrote later, "I quickly took advantage of the absence, or lack of circumspection, of the heirs, by taking the observations under my care, or perhaps usurping them." The result was Kepler's *Rudolphine Tables*, a compilation of the data from thirty years of Brahe's observations. To be fair, on his deathbed Brahe had urged Kepler to complete the tables; but Kepler did not frame the work according to any Tychonic hypothesis, as Brahe had hoped. Instead, Kepler used the data, which included calculations using logarithms he had developed himself, in predicting planetary positions. He was able to predict transits of the sun by Mercury and Venus, though he did not live long enough to witness them.

Kepler did not publish *Rudolphine Tables* until 1627, however, because the data he discovered constantly led him in new directions.

After Brahe's death, Kepler witnessed a nova, which later became known as "Kepler's nova," and he also experimented in optical theories. Though scientists and scholars view Kepler's optical work as minor in comparison with his accomplishments in astronomy and mathematics, the publication in 1611 of his book *Dioptrice* changed the course of optics.

In 1605, Kepler announced his first law, the law of ellipses, which held that the planets move in ellipses with the sun at one focus. Earth, Kepler asserted, is closest to the sun in January and farthest from it in July as it travels along its elliptical orbit. His second law, the law of equal areas, maintained that the farther a planet is from the sun, the more slowly it moves along its orbit. Kepler demonstrated this by arguing that an imaginary line connecting any planet to the sun must sweep over equal areas in equal intervals of time. He published both laws in 1609 in his book *New Astronomy* (*Astronomia Nova*).

Yet despite his status as Imperial Mathematician and as a distinguished scientist whom Galileo sought out for an opinion on his new telescopic discoveries, Kepler was unable to secure for himself a comfortable existence. Religious upheaval in Prague jeopardized his new homeland, and in 1611 his wife and his favorite son died. Kepler was permitted, under exemption, to return to Linz, and in 1613 he married Susanna Reuttinger, a twenty-four-year-old orphan who would bear him seven children, only two of whom would survive to adulthood. It was at this time that Kepler's mother was accused of witchcraft, and in the midst of his own personal turmoil he was forced to defend her against the charge in order to prevent her being burned at the stake. Katherine was imprisoned and tortured, but her son managed to obtain an acquittal and she was released.

Because of these distractions, Kepler's return to Linz was not a productive time initially. Distraught, he turned his attention away from tables and began working on *Harmonies of the World* (*Harmonice Mundi*), a passionate work which Max Caspar, in his biography of Kepler, described as "a great cosmic vision, woven out of science, poetry, philosophy, theology, mysticism." Kepler finished *Harmonies of the World* on May 27, 1618. In this series of five books, he extended his theory of harmony music, astrology, geometry, and astronomy. The series included his Kepler's third law of planetary motion, the law that would inspire Isaac Newton some sixty years later, which maintained that the cubes of mean distances of the planets of the sun are proportional to the squares of their periods of revolution. In short, Kepler discovered how planets orbited, and in so doing paved the way for Newton to discover why.

Kepler believed he had discovered God's logic in designing the universe, and he was unable to hide his ecstasy. In Book 5 of *Harmonies of the World* he wrote:

> *I dare frankly to confess that I have stolen the golden vessels of the Egyptians to build a tabernacle for my God far from the bounds of Egypt. If you pardon me, I shall rejoice; if you reproach me, I shall endure. The die is cast, and I am writing the book, to be read either now or by posterity, it matters not. It can wait a century for a reader, as God himself has waited six thousand years for a witness.*

The Thirty Years War, which began in 1618 and decimated the Austrian and Germanlands, forced Kepler to leave Linz in 1626. He eventually settled in the town of Sagan, in Silesia. There he tried to finish what might best be described as a science fiction novel, which he had dabbled at for years, at some expense to his mother during her trial for witchcraft. *Dream of the Moon* (*Somnium seu astronomia lunari*), which features an interview with a knowing "demon" who explains how the protagonist could travel to the moon, was uncovered and presented as evidence during Katherine's trial. Kepler spent considerable energy defending the work as pure fiction and the demon as a mere literary device. The book was unique in that it was not only ahead of its time in terms of fantasy but also a treatise supporting Copernican theory.

In 1630, at the age of fifty-nine, Kepler once again found himself in financial straits. He set out for Regensburg, where he hoped to collect interest on some bonds in his possession as well as some money he was owed. However, a few days after his arrival he developed a fever and died on November 15. Though he never achieved the mass renown of Galileo, Kepler produced a body of work that was extraordinarily useful to professional astronomers, like Newton, who immersed themselves in the details and accuracy of Kepler's science. Johannes Kepler was a man who preferred aesthetic harmony and order, and all that he discovered was inextricably linked with his vision of God. His epitaph, which he himself composed, reads: "I used to measure the heavens; now I shall measure the shadows of the earth. Although my soul was from heaven, the shadow of my body lies here."

Harmonies

of the

World

BOOK FIVE

Concerning the very perfect harmony of the celestial movements, and the genesis of eccentricities and the semidiameters, and as periodic times from the same.

After the model of the most correct astronomical doctrine of today, and the hypothesis not only of Copernicus but also of Tycho Brahe, whereof either hypotheses are today publicly accepted as most true, and the Ptolemaic as outmoded.

I commence a sacred discourse, a most true hymn to God the Founder, and I judge it to be piety, not to sacrifice many hecatombs of bulls to Him and to burn incense of innumerable perfumes and cassia, but first to learn myself, and afterwards to teach others too, how great He is in wisdom, how great in power, and of what sort in goodness. For to wish to adorn in every way possible the things that should receive adornment and to envy no thing its goods—this I put down as the sign of the greatest goodness, and in this respect I praise Him as good that in the heights of His wisdom He finds everything whereby each

1

thing may be adorned to the utmost and that He can do by His unconquerable power all that He has decreed.

Galen, *on the Use of Parts*. Book III

Proem

[268] As regards that which I prophesied two and twenty years ago (especially that the five regular solids are found between the celestial spheres), as regards that of which I was firmly persuaded in my own mind before I had seen Ptolemy's *Harmonies*, as regards that which I promised my friends in the title of this fifth book before I was sure of the thing itself, that which, sixteen years ago, in a published statement, I insisted must be investigated, for the sake of which I spent the best part of my life in astronomical speculations, visited Tycho Brahe, [269] and took up residence at Prague: finally, as God the Best and Greatest, Who had inspired my mind and aroused my great desire, prolonged my life and strength of mind and furnished the other means through the liberality of the two Emperors and the nobles of this province of Austria-on-the-Anisana: after I had discharged my astronomical duties as much as sufficed, finally, I say, I brought it to light and found it to be truer than I had even hoped, and I discovered among the celestial movements the full nature of harmony, in its due measure, together with all its parts unfolded in Book III—not in that mode wherein I had conceived it in my mind (this is not last in my joy) but in a very different mode which is also very excellent and very perfect. There took place in this intervening time, wherein the very laborious reconstruction of the movements held me in suspense, an extraordinary augmentation of my desire and incentive for the job, a reading of the *Harmonies* of Ptolemy, which had been sent to me in manuscript by John George Herward, Chancellor of Bavaria, a very distinguished man and of a nature to advance philosophy and every type of learning. There, beyond my expectations and with the greatest wonder, I found approximately the whole third book given over to the same consideration of celestial harmony, fifteen hundred years ago. But indeed astronomy was far from being of age as yet; and Ptolemy, in an unfortunate attempt, could make others subject to despair, as being one who, like Scipio in Cicero, seemed to have recited a pleasant Pythagorean dream rather than to have aided philosophy. But both the crudeness of the ancient philosophy and this exact agreement in our meditations, down to the last hair, over an interval of fifteen centuries, greatly strengthened me in getting on with the job. For what need is there of many men? The very nature of things, in order to reveal herself to mankind, was at work in the different interpreters of different ages, and was the finger of God—to use the Hebrew expression; and here, in the minds of two men, who had wholly given themselves up to the contemplation of nature, there was the same conception as to the configuration of the world, although neither had

been the other's guide in taking this route. But now since the first light eight months ago, since broad day three months ago, and since the sun of my wonderful speculation has shone fully a very few days ago: nothing holds me back. I am free to give myself up to the sacred madness, I am free to taunt mortals with the frank confession that I am stealing the golden vessels of the Egyptians, in order to build of them a temple for my God, far from the territory of Egypt. If you pardon me, I shall rejoice; if you are enraged, I shall bear up. The die is cast, and I am writing the book—whether to be read by my contemporaries or by posterity matters not. Let it await its reader for a hundred years, if God Himself has been ready for His contemplator for six thousand years.

The chapters of this book are as follows:

1. Concerning the five regular solid figures.
2. On the kinship between them and the harmonic ratios.
3. Summary of astronomical doctrine necessary for speculation into the celestial harmonies.
4. In what things pertaining to the planetary movements the simple consonances have been expressed and that all those consonances which are present in song are found in the heavens.
5. That the clefs of the musical scale, or pitches of the system, and the genera of consonances, the major and the minor, are expressed in certain movements.
6. That the single musical Tones or Modes are somehow expressed by the single planets.
7. That the counterpoints or universal harmonies of all the planets can exist and be different from one another.
8. That four kinds of voice are expressed in the planets: soprano, contralto, tenor, and bass.
9. Demonstration that in order to secure this harmonic arrangement, those very planetary eccentricities which any planet has as its own, and no others, had to be set up.
10. Epilogue concerning the sun, by way of very fertile conjectures.

[270] Before taking up these questions, it is my wish to impress upon my readers the very exhortation of Timaeus, a pagan philosopher, who was going to speak on the same things: it should be learned by Christians with the greatest admiration, and shame too, if they do not imitate him: Ἀλλ' ὦ Σώκρατες, τοῦτο γε δὴ πντὲς, ὅσοι καὶ κατὰ βραχὺ σωφροσυνης μετέχουσιν, ἐπὶ πασῇ ὁρμῇ καὶ σμίκρου καὶ μεγάλου πράγματσο θεὸν ἀει που καλοῦσιν. ἡμᾶς δὲ τοὺς περὶ τοῦ πάντος λόγους ποιεῖσθαι πη μέλλοντᾶς..., εἰ μὴ πανταπασι παραλλάτομεν, ἀνάγκη θεοὺς τε καί θεὰς ἐπικαλουμένους εὔχεσθαι πάντα, κατὰ νοῦν ἐκείνοις μέν μάλιστα, ἐπομένως δέ ἡμῖν εἰπεῖν. *For truly, Socrates, since all who*

have the least particle of intelligence always invoke God whenever they enter upon any business, whether light or arduous; so too, unless we have clearly strayed away from all sound reason, we who intend to have a discussion concerning the universe must of necessity make our sacred wishes and pray to the Gods and Goddesses with one mind that we may say such things as will please and be acceptable to them in especial and, secondly, to you too.

1. CONCERNING THE FIVE REGULAR SOLID FIGURES

[271] It has been said in the second book how the regular plane figures are fitted together to form solids; there we spoke of the five regular solids, among others, on account of the plane figures. Nevertheless their number, five, was there demonstrated; and it was added why they were designated by the Platonists as the figures of the world, and to what element any solid was compared on account of what property. But now, in the anteroom of this book, I must speak again concerning these figures, on their own account, not on account of the planes, as much as suffices for the celestial harmonies; the reader will find the rest in the *Epitome of Astronomy*, Volume II, Book IV.

Accordingly, from the *Mysterium Cosmographicum*, let me here briefly inculcate the order of the five solids in the world, whereof three are primary and two secondary. For the *cube* (1) is the outmost and the most spacious, because firstborn and having the nature (*rationem*) of a *whole*, in the very form of its generation. There follows the *tetrahedron* (2), as if made a *part*, by cutting up the cube; nevertheless it is primary too, with a solid trilinear angle, like the cube. Within the tetrahedron is the *dodecahedron* (3), the last of primary figures, namely, like a solid composed of parts of a cube and similar parts of a tetrahedron, *i.e.*, of irregular tetrahedrons, wherewith the cube inside is roofed over. Next in order is the *icosahedron* (4) on account of its similarity, the last of the secondary figures and having a plurilinear solid angle. The *octahedron* (5) is inmost, which is similar to the cube and the first of the secondary figures and to which as inscriptile the first place is due, just as the first outside place is due to the cube as circumscriptile.

[272] However, there are as it were two noteworthy weddings of these figures, made from different classes: the males, the cube and the dodecahedron, among the primary; the females, the octahedron and the icosahedron, among the secondary, to which is added one as it were bachelor or hermaphrodite, the tetrahedron, because it is inscribed in itself, just as those female solids are inscribed in the males and are as it were subject to them, and have the signs of the feminine sex, opposite the masculine, namely, angles opposite planes. Moreover, just as the tetrahedron is the element, bowels, and as it were rib of the male cube, so the feminine octahedron is the element and part of the tetrahedron in another way; and thus the tetrahedron mediates in this marriage.

The main difference in these wedlocks or family relationships consists in the following: the ratio of the cube is *rational.* For the tetrahedron is one third of the body of the cube, and the octahedron half of the tetrahedron, one sixth of the cube; while the ratio of the dodecahedron's wedding is *irrational (ineffabilis)* but *divine.*

The union of these two words commands the reader to be careful as to their significance. For the word *ineffabilis* here does not of itself denote any nobility, as elsewhere in theology and divine things, but denotes an inferior condition. For in geometry, as was said in the first book, there are many irrationals, which do not on that account participate in a divine proportion too. But you must look in the first book for what the divine ratio, or rather the divine section, is. For in other proportions there are four terms present; and three, in a continued proportion; but the divine requires a single relation of terms outside of that of the proportion itself, namely in such fashion that the two lesser terms, as parts make up the greater term, as a whole. Therefore, as much as is taken away from this wedding of the dodecahedron on account of its employing an irrational proportion,

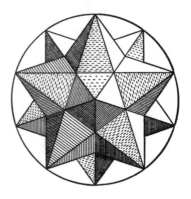

is added to it conversely, because its irrationality approaches the divine. This wedding also comprehends the solid star too, the generation whereof arises from the continuation of five planes of the dodecahedron till they all meet in a single point. See its generation in Book II.

Lastly, we must note the ratio of the spheres circumscribed around them to those inscribed in them: in the case of the tetrahedron it is rational, 100,000 : 33,333 or 3 : 1; in the wedding of the cube it is irrational, but the radius of the inscribed sphere is rational in square, and is itself the square root of one third the square on the radius (of the circumscribed sphere), namely 100,000 : 57,735; in the wedding of the dodecahedron, clearly irrational, 100,000 : 79,465; in the case of the star, 100,000 : 52,573, half the side of the icosahedron or half the distance between two rays.

2. ON THE KINSHIP BETWEEN THE HARMONIC RATIOS AND THE FIVE REGULAR FIGURES

[273] This kinship (*cognatio*) is various and manifold; but there are four degrees of kinship. For either the sign of kinship is taken from the outward form alone which the figures have, or else ratios which are the same as the harmonic arise in the construction of the side, or result from the figures already constructed, taken simply or together; or, lastly, they are either equal to or approximate the ratios of the spheres of the figure.

In the first degree, the ratios, where the character or greater term is 3, have kinship with the triangular plane of the tetrahedron, octahedron, and icosahedron; but where the greater term is 4, with the square plane of the cube; where 5, with the pentagonal plane of the dodecahedron. This similitude on the part of the plane can also be extended to the smaller term of the ratio, so that wherever the number 3 is found as one term of the continued doubles, that ratio is held to be akin to the three figures first named: for example, 1 : 3 and 2 : 3 and 4 : 3 and 8 : 3, et cetera; but where the number is 5, that ratio is absolutely assigned to the wedding of the dodecahedron: for example, 2 : 5 and 4 : 5 and 8 : 5, and thus 3 : 5 and 3 : 10 and 6 : 5 and 12 : 5 and 24 : 5. The kinship will be less probable if the sum of the terms expresses this similitude, as in 2 : 3 the sum of the terms is equal to 5, as if to say that 2 : 3 is akin to the dodecahedron. The kinship on account of the outward form of the solid angle is similar: the solid angle is trilinear among the primary figures, quadrilinear in the octahedron, and quinquelinear in the icosahedron. And so if one term of the ratio participates in the number 3, the ratio will be connected with the primary bodies; but if in the number 4, with the octahedron; and finally, if in the number 5, with the icosahedron. But in the feminine solids this kinship is more apparent, because the characteristic figure latent within follows upon the form of the angle: the tetragon in the octahedron, the pentagon in the icosahedron; and so 3 : 5 would go to the sectioned icosahedron for both reasons.

The second degree of kinship, which is genetic, is to be conceived as follows: First, some harmonic ratios of numbers are akin to one wedding or family, namely, perfect ratios to the single family of the cube; conversely, there is the ratio which is never fully expressed in numbers and cannot be demonstrated by numbers in any other way, except by a long series of numbers gradually approaching it: this ratio is called *divine*, when it is perfect, and it rules in various ways throughout the dodecahedral wedding. Accordingly, the following consonances begin to shadow forth that ratio: 1 : 2 and 2 : 3 and 2 : 3 and 5 : 8. For it exists most imperfectly in 1 : 2, more perfectly in 5 : 8, and still more perfectly if we add 5 and 8 to make 13 and take 8 as the numerator, if this ratio has not stopped being harmonic.

Further, in constructing the side of the figure, the diameter of the globe must be cut; and the octahedron demands its bisection, the cube and the tetrahedron its trisection, the dodecahedral wedding its quinquesection. Accordingly, the ratios between the figures are distributed according to the numbers which express those ratios. But the square on the diameter is cut too, or the square on the side of the figure is formed from a fixed part of the diameter. And then the squares on the sides are compared with the square on the diameter, and they constitute the following ratios: in the cube 1 : 3, in the tetrahedron 2 : 3, in the octahedron 1 : 2. Wherefore, if the two ratios are put together, the cubic and the tetrahedral will give 1 : 2; the cubic and the octahedral,

2 : 3; the octahedral and the tetrahedral, 3 : 4. The sides in the dodecahedral wedding are irrational.

Thirdly, the harmonic ratios follow in various ways upon the already constructed figures. For either the number of the sides of the plane is compared with the number of lines in the total figure; [274] and the following ratios arise: in the cube, 4 : 12 or 1 : 3; in the tetrahedron 3 : 6 or 1 : 2; in the octahedron 3 : 12 or 1 : 4; in the dodecahedron 5 : 30 or 1 : 6; in the icosahedron 3 : 30 or 1 : 10. Or else the number of sides of the plane is compared with the number of planes; then the cube gives 4 : 6 or 2 : 3, the tetrahedron 3 : 4, the octahedron 3 : 8, the dodecahedron 5 : 12, the icosahedron 3 : 20. Or else the number of sides or angles of the plane is compared with the number of solid angles, and the cube gives 4 : 8 or 1 : 2, the tetrahedron 3 : 4, the octahedron 3 : 6 or 1 : 2, the dodecahedron with its consort 5 : 20 or 3 : 12 (i.e., 1 : 4). Or else the number of planes is compared with the number of solid angles, and the cubic wedding gives 6 : 8 or 3 : 4, the tetrahedron the ratio of equality, the dodecahedral wedding 12 : 20 or 3 : 5. Or else the number of all the sides is compared with the number of the solid angles, and the cube gives 8 : 12 or 2 : 3, the tetrahedron 4 : 6 or 2 : 3, and the octahedron 6 : 12 or 1 : 2, the dodecahedron 20 : 30 or 2 : 3, the icosahedron 12 : 30 or 2 : 5.

Moreover, the bodies too are compared with one another, if the tetrahedron is stowed away in the cube, the octahedron in the tetrahedron and cube, by geometrical inscription. The tetrahedron is one third of the cube, the octahedron half of the tetrahedron, one sixth of the cube, just as the octahedron, which is inscribed in the globe, is one sixth of the cube which circumscribes the globe. The ratios of the remaining bodies are irrational.

The fourth species or degree of kinship is more proper to this work: the ratio of the spheres inscribed in the figures to the spheres circumscribing them is sought, and what harmonic ratios approximate them is calculated. For only in the tetrahedron is the diameter of the inscribed sphere rational, namely, one third of the circumscribed sphere. But in the cubic wedding the ratio, which is single there, is as lines which are rational only in square. For the diameter of the inscribed sphere is to the diameter of the circumscribed sphere as the square root of the ratio 1 : 3. And if you compare the ratios with one another, the ratio of the tetrahedral spheres is the square of the ratio of the cubic spheres. In the dodecahedral wedding there is again a single ratio, but an irrational one, slightly greater than 4 : 5. Therefore the ratio of the spheres of the cube and octahedron is approximated by the following consonances: 1 : 2, as proximately greater, and 3 : 5, as proximately smaller. But the ratio of the dodecahedral spheres is approximated by the consonances 4 : 5 and 5 : 6, as proximately smaller, and 3 : 4 and 5 : 8, as proximately greater.

But if for certain reasons 1 : 2 and 1 : 3 are arrogated to the cube, the ratio of the spheres of the cube will be to the ratio of the spheres of the tetrahedron as the consonances 1 : 2 and 1 : 3, which have been ascribed to the cube, are to 1 : 4 and 1 : 9, which are to be assigned to the tetrahedron, if this proportion is to be used. For these ratios, too, are as the squares of those consonances. And because 1 : 9 is not harmonic, 1 : 8 the proximate ratio takes its place in the tetrahedron. But by this proportion approximately 4 : 5 and 3 : 4 will go with the dodecahedral wedding. For as the ratio of the spheres of the cube is approximately the cube of the ratio of the dodecahedral, so too the cubic consonances 1 : 2 and 2 : 3 are approximately the cubes of the consonances 4 : 5 and 3 : 4. For 4 : 5 cubed is 64 : 125, and 1 : 2 is 64 : 128. So 3 : 4 cubed is 27 : 64, and 1 : 3 is 27 : 81.

3. A SUMMARY OF ASTRONOMICAL DOCTRINE NECESSARY FOR SPECULATION INTO THE CELESTIAL HARMONIES

First of all, my readers should know that the ancient astronomical hypotheses of Ptolemy, in the fashion in which they have been unfolded in the *Theoricae* of Peurbach and by the other writers of epitomes, are to be completely removed from this discussion and cast out of [275] the mind. For they do not convey the true lay out of the bodies of the world and the polity of the movements.

Although I cannot do otherwise than to put solely Copernicus' opinion concerning the world in the place of those hypotheses and, if that were possible, to persuade everyone of it; but because the thing is still new among the mass of the intelligentsia (*apud vulgus studiosorum*), and the doctrine that the Earth is one of the planets and moves among the stars around a motionless sun sounds very absurd to the ears of most of them: therefore those who are shocked by the unfamiliarity of this opinion should know that these harmonical speculations are possible even with the hypotheses of Tycho Brahe—because that author holds, in common with Copernicus, everything else which pertains to the lay out of the bodies and the tempering of the movements, and transfers solely the Copernican annual movement of the Earth to the whole system of planetary spheres and to the sun, which occupies the centre of that system, in the opinion of both authors. For after this transference of movement it is nevertheless true that in Brahe the Earth occupies at any time the same place that Copernicus gives it, if not in the very vast and measureless region of the fixed stars, at least in the system of the planetary world. And accordingly, just as he who draws a circle on paper makes the writing-foot of the compass revolve, while he who fastens the paper or tablet to a turning lathe draws the same circle on the revolving tablet with the foot of the compass or stylus motionless; so too, in the case of Copernicus the Earth, by the real movement of its body, measures out a circle revolving midway between the circle of Mars on the

8

outside and that of Venus on the inside; but in the case of Tycho Brahe the whole planetary system (wherein among the rest the circles of Mars and Venus are found) revolves like a tablet on a lathe and applies to the motionless Earth, or to the stylus on the lathe, the midspace between the circles of Mars and Venus; and it comes about from this movement of the system that the Earth within it, although remaining motionless, marks out the same circle around the sun and midway between Mars and Venus, which in Copernicus it marks out by the real movement of its body while the system is at rest. Therefore, since harmonic speculation considers the eccentric movements of the planets, as if seen from the sun, you may easily understand that if any observer were stationed on a sun as much in motion as you please, nevertheless for him the Earth, although at rest (as a concession to Brahe), would seem to describe the annual circle midway between the planets and in an intermediate length of time. Wherefore, if there is any man of such feeble wit that he cannot grasp the movement of the Earth among the stars, nevertheless he can take pleasure in the most excellent spectacle of this most divine construction, if he applies to their image in the sun whatever he hears concerning the daily movements of the Earth in its eccentric—such an image as Tycho Brahe exhibits, with the Earth at rest.

And nevertheless the followers of the true Samian philosophy have no just cause to be jealous of sharing this delightful speculation with such persons, because their joy will be in many ways more perfect, as due to the consummate perfection of speculation, if they have accepted the immobility of the sun and the movement of the Earth.

Firstly [I], therefore, let my readers grasp that today it is absolutely certain among all astronomers that all the planets revolve around the sun, with the exception of the moon, which alone has the Earth as its centre: the magnitude of the moon's sphere or orbit is not great enough for it to be delineated in this diagram in a just ratio to the rest. Therefore, to the other five planets, a sixth, the Earth, is added, which traces a sixth circle around the sun, whether by its own proper movement with the sun at rest, or motionless itself and with the whole planetary system revolving.

Secondly [II]: It is also certain that all the planets are eccentric, *i.e.*, they change their distances from the sun, in such fashion that in one part of their circle they become farthest away from the sun, [276] and in the opposite part they come nearest to the sun. In the accompanying diagram three circles apiece have been drawn for the single planets: none of them indicate the eccentric route of the planet itself; but the mean circle, such as *BE* in the case of Mars, is equal to the eccentric orbit, with respect to its longer diameter. But the orbit itself, such as *AD*, touches *AF*, the upper of the three, in one place *A*, and the lower circle *CD*, in the opposite place *D*. The circle *GH* made with dots and described through the centre of the sun indicates the route of the sun according to Tycho Brahe. And if the sun moves on this route, then absolutely all the points in this whole planetary system here depicted advance upon an equal route, each upon his own. And with one point of it (namely, the centre of the sun) stationed at one point of its circle, as here at the lowest, absolutely each and every point of the system will be stationed at the lowest part of its circle. However, on account of the smallness of the space the three circles of Venus unite in one, contrary to my intention.

Thirdly [III]: Let the reader recall from my *Mysterium Cosmographicum*, which I published twenty-two years ago, that the number of the planets or circular routes around the sun was taken by the very wise Founder from the five regular solids, concerning which Euclid, so many ages ago, wrote his book which is called the *Elements* in that it is built up out of a series of propositions. But it has been made clear in the second book of this work that there cannot be more regular

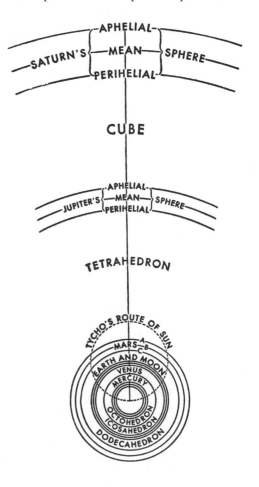

bodies, *i.e.*, that regular plane figures cannot fit together in a solid more than five times.

Fourthly [IV]: As regards the ratio of the planetary orbits, the ratio between two neighbouring planetary orbits is always of such a magnitude that it is easily apparent that each and every one of them approaches the single ratio of the spheres of one of the five regular solids, namely, that of the sphere circumscribing to the sphere inscribed in the figure. Nevertheless it is not wholly equal, as I once dared to promise concerning the final perfection of astronomy. For, after completing the demonstration of the intervals from Brahe's observations, I discovered the following: if the angles of the cube [277] are applied to the inmost circle of Saturn, the centres of the planes are approximately tangent to the middle circle of Jupiter; and if the angles of the tetrahedron are placed against the inmost circle of Jupiter, the centres of the planes of the tetrahedron are approximately tangent to the outmost circle of Mars; thus if the angles of the octahedron are placed against any circle of Venus (for the total interval between the three has been very much reduced), the centres of the planes of the octahedron penetrate and descend deeply within the outmost circle of Mercury, but nonetheless do not reach as far as the middle circle of Mercury; and finally, closest of all to the ratios of the dodecahedral and icosahedral spheres—which ratios are equal to one another—are the ratios or intervals between the circles of Mars and the Earth, and the Earth and Venus; and those intervals are similarly equal, if we compute from the inmost circle of Mars to the middle circle of the Earth, but from the middle circle of the Earth to the middle circle of Venus. For the middle distance of the Earth is a mean proportional between the least distance of Mars and the middle distance of Venus. However, these two ratios between the planetary circles are still greater than the ratios of those two pairs of spheres in the figures, in such fashion that the centres of the dodecahedral planes are not tangent to the outmost circle of the Earth, and the centres of the icosahedral planes are not tangent to the outmost circle of Venus; nor, however, can this gap be filled by the semidiameter of the lunar sphere, by adding it, on the upper side, to the greatest distance of the Earth and subtracting it, on the lower, from the least distance of the same. But I find a certain other ratio of figures—namely, if I take the augmented dodecahedron, to which I have given the name of echinus, (as being fashioned from twelve quinquangular stars and thereby very close to the five regular solids), if I take it, I say, and place its twelve points in the inmost circle of Mars, then the sides of the pentagons, which are the bases of the single rays or points, touch the middle circle of Venus. In short: the cube and the octahedron, which are consorts, do not penetrate their planetary spheres at all; the dodecahedron and the icosahedron, which are consorts, do not wholly reach to theirs, the tetrahedron exactly touches both: in the first case there is falling short; in the second, excess; and in the third, equality, with respect to the planetary intervals.

Wherefore it is clear that the very ratios of the planetary intervals from the sun have not been taken from the regular solids alone. For the Creator, who is the very source of geometry and, as Plato wrote, "practices eternal geometry," does not stray from his own archetype. And indeed that very thing could be inferred from the fact that all the planets change their intervals throughout fixed periods of time, in such fashion that each has two marked intervals from the sun, a greatest and a least; and a fourfold comparison of the intervals from the sun is possible between two planets: the comparison can be made between either the greatest, or the least, or the contrary intervals most remote from one another, or the contrary intervals nearest together. In this way the comparisons made two by two between neighbouring planets are twenty in number, although on the contrary there are only five regular solids. But it is consonant that if the Creator had any concern for the ratio of the spheres in general, He would also have had concern for the ratio which exists between the varying intervals of the single planets specifically and that the concern is the same in both cases and the one is bound up with the other. If we ponder that, we will comprehend that for setting up the diameters and eccentricities conjointly, there is need of more principles, outside of the five regular solids.

Fifthly [V]: To arrive at the movements between which the consonances have been set up, once more I impress upon the reader that in the *Commentaries on Mars* I have demonstrated from the sure observations of Brahe that daily arcs, which are equal in one and the same eccentric circle, are not traversed with equal speed; but that these differing *delays in equal parts of the eccentric observe the ratio of their distances from the sun,* the source of movement; and conversely, that if equal times are assumed, namely, one natural day in both cases, the corresponding *true diurnal arcs* [278] *of one eccentric orbit have to one another the ratio which is the inverse of the ratio of the two distances from the sun.* Moreover, I demonstrated at the same time that *the planetary orbit is elliptical and the sun, the source of movement, is at one of the foci of this ellipse; and so, when the planet has completed a quarter of its total circuit from its aphelion, then it is exactly at its mean distance from the sun, midway between its greatest distance at the aphelion and its least at the perihelion.* But from these two axioms it results *that the diurnal mean movement of the planet in its eccentric is the same as the true diurnal arc of its eccentric at those moments wherein the planet is at the end of the quadrant of the eccentric measured from the aphelion, although that true quadrant appears still smaller than the just quadrant.* Furthermore, it follows *that the sum of any two true diurnal eccentric arcs, one of which is at the same distance from the aphelion that the other is from the perihelion, is equal to the sum of the two mean diurnal arcs.* And as a consequence, *since the ratio of circles is the same as that of the diameters, the ratio of one mean diurnal arc to the sum of all the mean and equal arcs in the total circuit is the same as the ratio of the mean diurnal arc to the sum of all the true eccentric arcs, which are the same in number but unequal to one another.* And those things should first be known concerning the true diurnal arcs of the eccentric and the true movements, so that by means of them we may understand the movements which would be apparent if we were to suppose an eye at the sun.

Sixthly [VI]: But as regards the arcs which are apparent, as it were, from the sun, it is known even from the ancient astronomy that, among true movements which are equal to one another, that movement which is farther distant from the centre of the world (as being at the aphelion) will appear smaller to a beholder at that centre, but the movement which is nearer (as being at the perihelion) will similarly appear greater. Therefore, since moreover the true diurnal arcs at the near distance are still greater, on account of the faster movement, and still smaller at the distant aphelion, on account of the slowness of the movement, I demonstrated in the *Commentaries on Mars* that *the ratio of the apparent diurnal arcs of one eccentric circle is fairly exactly the inverse ratio of the squares of their distances from the sun.* For example, if the planet one day when it is at a distance from the sun of 10 parts, in any measure whatsoever, but on the opposite day, when it is at the perihelion, of 9 similar parts: it is certain that from the sun its apparent progress at the aphelion will be to its apparent progress at the perihelion, as 81 : 100.

But that is true with these provisos: First, that the eccentric arcs should not be great, lest they partake of distinct distances which are very different—*i.e.,* lest the distances of their termini from the apsides cause a perceptible variation; second, that the eccentricity should not be very great, for the greater its eccentricity (*viz.,* the greater the arc becomes) the more the angle of its apparent movement increases beyond the measure of its approach to the sun, by Theorem 8 of Euclid's *Optics;* none the less in small arcs even a great distance is of no moment, as I have remarked in my *Optics,* Chapter 11. But there is another reason why I make that admonition. For the eccentric arcs around the mean anomalies are viewed obliquely from the centre of the sun. This obliquity subtracts from the magnitude of the apparent movement, since conversely the arcs around the apsides are presented directly to an eye stationed as it were at the sun. Therefore, when the eccentricity is very great, then the eccentricity takes away perceptibly from the ratio of the movements; if without any diminution we apply the mean diurnal movement to the mean distance, as if at the mean distance, it would appear to have the same magnitude which it does have— as will be apparent below in the case of Mercury. All these things are treated at greater length in Book V of the *Epitome of Copernican Astronomy;* but they have been mentioned here too because they have to do with the very terms of the celestial consonances, considered in themselves singly and separately.

Seventhly [VII]: If by chance anyone runs into those diurnal movements which are apparent [279] to those gazing not as it were from the sun but from the Earth, with which movements Book VI of the *Epitome of Copernican Astronomy* deals, he should know that their rationale is plainly not considered in this business. Nor should it be, since the Earth is not the source of the planetary movements, nor can it be, since with respect to deception of sight they degenerate not only into mere quiet or apparent stations but even into retrogradation, in which way a whole infinity of ratios is assigned to all the planets, simultaneously and equally. Therefore, in order that we may hold for certain what sort of ratios of their own are constituted by the single real eccentric orbits (although these too are still apparent, as it were to one looking from the sun, the source of movement), first we must remove from those movements of their own this image of the adventitious annual movement common to all five, whether it arises from the movement of the Earth itself, according to Copernicus, or from the annual movement of the total system, according to Tycho Brahe, and the winnowed movements proper to each planet are to be presented to sight.

Eighthly [VIII]: So far we have dealt with the different delays or arcs of one and the same planet. Now we must also deal with the comparison of the movements of two planets. Here take note of the definitions of the terms which will be necessary for us. We give the name of *nearest apsides* of two planets to the perihelion of the upper and the aphelion of the lower, notwithstanding that they tend not towards the same region of the world but towards distinct and perhaps contrary regions. By *extreme movements* understand the slowest and the fastest of the whole planetary circuit; by *converging or converse extreme movements*, those which are at the nearest apsides of two planets— namely, at the perihelion of the upper planet and the aphelion of the lower; by *diverging or diverse*, those at the opposite apsides—namely, the aphelion of the upper and the perihelion of the lower. Therefore again, a certain part of my *Mysterium Cosmographicum*, which was suspended twenty-two years ago, because it was not yet clear, is to be completed and herein inserted. For after finding the true intervals of the spheres by the observations of Tycho Brahe and continuous labour and much time, at last, at last the right ratio of the periodic times to the spheres

> *though it was late, looked to the unskilled man,*
> *yet looked to him, and, after much time, came,*

and, if you want the exact time, was conceived mentally on the 8th of March in this year One Thousand Six Hundred and Eighteen but unfelicitously submitted to calculation and rejected as false, finally, summoned back on the 15th of May, with a fresh assault undertaken, outfought the darkness of my mind by the great proof afforded by my labor of seventeen years on Brahe's observations and meditation upon it uniting in one concord, in such fashion that I first believed I was dreaming and was presupposing the object of my search among the principles. But it is absolutely certain and exact that *the ratio which exists between the periodic times of any two planets is precisely the ratio of the* $^3/_2$*th power of the mean distances,* i.e., *of the spheres themselves,* provided, however, that the arithmetic mean between both diameters of the elliptic orbit be slightly less than the longer diameter. And so if any one take the period, say, of the Earth, which is one year, and the period of Saturn, which is thirty years, and extract the cube roots of this ratio and then square the ensuing ratio by squaring the cube roots, he will have as his numerical products the most just ratio of the distances of the Earth and Saturn from the sun.[1] For the cube root of 1 is 1, and the square of it is 1; and the cube root of 30 is greater than 3, and therefore the square of it is greater than 9. And Saturn, at

---

1. For in the *Commentaries on Mars*, chapter 48, page 232, I have proved that this Arithmetic mean is either the diameter of the circle which is equal in length to the elliptic orbit, or else is very slightly less.

its mean distance from the sun, is slightly higher [280] than nine times the mean distance of the Earth from the sun. Further on, in Chapter 9, the use of this theorem will be necessary for the demonstration of the eccentricities.

Ninthly [IX]: If now you wish to measure with the same yardstick, so to speak, the true daily journeys of each planet through the ether, two ratios are to be compounded—the ratio of the true (not the apparent) diurnal arcs of the eccentric, and the ratio of the mean intervals of each planet from the sun (because that is the same as the ratio of the amplitude of the spheres), *i.e., the true diurnal arc of each planet is to be multiplied by the semidiameter of its sphere:* the products will be numbers fitted for investigating whether or not those journeys are in harmonic ratios.

Tenthly [X]: In order that you may truly know how great any one of these diurnal journeys appears to be to an eye stationed as it were at the sun, although this same thing can be got immediately from the astronomy, nevertheless it will also be manifest if you multiply the ratio of the journeys by the inverse ratio not of the mean, but of the true intervals which exist at any position on the eccentrics: *multiply the journey of the upper by the interval of the lower planet from the sun, and conversely multiply the journey of the lower by the interval of the upper from the sun.*

Eleventhly [XI]: And in the same way, if the apparent movements are given, at the aphelion of the one and at the perihelion of the other, or conversely or alternately, the ratios of the distances of the aphelion of the one to the perihelion of the other may be elicited. But where the mean movements must be known first, *viz.,* the inverse ratio of the periodic times, wherefrom the ratio of the spheres is elicited by Article VIII above: then *if the mean proportional between the apparent movement of either one of its mean movement be taken, this mean proportional is to the semidiameter of its sphere* (which is already known) *as the mean movement is to the distance or interval sought.* Let the periodic times of two planets be 27 and 8. Therefore the ratio of the mean diurnal movement of the one to the other is 8 : 27. Therefore the semidiameters of their spheres will be as 9 to 4. For the cube root of 27 is 3, that of 8 is 2, and the squares of these roots, 3 and 2, are 9 and 4. Now let the apparent aphelial movement of the one be 2 and the perihelial movement of the other $33^1/3$. The mean proportionals between the mean movements 8 and 27 and these apparent ones will be 4 and 30. Therefore if the mean proportional 4 gives the mean distance of 9 to the planet, then the mean movement of 8 gives an aphelial distance 18, which corresponds to the apparent movement 2; and if the other mean proportional 30 gives the other planet a mean distance of 4, then its mean movement of 27 will give it a perihelial interval of $3^3/5$. I say, therefore, that the aphelial distance of the former is to the perihelial distance of the latter as 18 to $3^3/5$. Hence it is clear that if the consonances between the extreme movements of two planets

15

are found and the periodic times are established for both, the extreme and the mean distances are necessarily given, wherefore also the eccentricities.

Twelfthly [XII]: It is also possible, from the different extreme movements of one and the same planet, to find the *mean movement*. The mean movement is not exactly the arithmetic mean between the extreme movements, nor exactly the geometric mean, but it is as much less than the geometric mean as the geometric mean is less than the (arithmetic) mean between both means. Let the two extreme movements be 8 and 10: the mean movement will be less than 9, and also less than the square root of 80 by half the difference between 9 and the square root of 80. In this way, if the aphelial movement is 20 and the perihelial 24, the mean movement will be less than 22, even less than the square root of 480 by half the difference between that root and 22. There is use for this theorem in what follows.

[281] Thirteenthly [XIII]: From the foregoing the following proposition is demonstrated, which is going to be very necessary for us: Just as the ratio of the mean movements of two planets is the inverse ratio of the $3/2$th powers of the spheres, so the ratio of two apparent converging extreme movements always falls short of the ratio of the $3/2$th powers of the intervals corresponding to those extreme movements; and in what ratio the product of the two ratios of the corresponding intervals to the two mean intervals or to the semidiameters of the two spheres falls short of the ratio of the square roots of the spheres, in that ratio does the ratio of the two extreme converging movements exceed the ratio of the corresponding intervals; but if that compound ratio were to exceed the ratio of the square roots of the spheres, then the ratio of the converging movements would be less than the ratio of their intervals.[1]

Let the ratio of the spheres be $DH : AE$; let the ratio of the mean movements be $HI : EM$, the $3/2$th power of the inverse of the former. Let the least interval of the sphere of the first be CG; and the greatest interval of the sphere of the second be BF; and first let $DH : CG$ comp. $BF : AE$ be smaller than the $1/2$th power of $DH : AE$. And let GH be the apparent perihelial movement of the upper planet, and FL the aphelial of the lower, so that they are converging extreme movements.

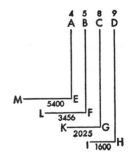

I say that

$$GK : FL = BF : CG$$
$$BF^{3/2} : CG^{3/2}.$$

For

$$HI : GK = CG^2 : DH^2$$

---

1. Kepler always measures the magnitude of a ratio from the greater term to the smaller, rather than from the antecedent to the consequent, as we do today. For example, as Kepler speaks, 2 : 3 is the same as 3 : 2, and 3 : 4 is greater than 7 : 8.—C. G. Wallis.

and
$$FL : EM = AE^2 : BF^2.$$
Hence
$$HI : GK \text{ comp. } FL : EM = CG^2 : DH^2 \text{ comp. } AE^2 : BF^2.$$
But
$$CG : DH \text{ comp. } AE : BF < AE^{1/2} : DH^{1/2}$$
by a fixed ratio of defect, as was assumed. Therefore too
$$HI : GK \text{ comp. } FL : EM \quad AE^{2/2} : DH^{2/2}$$
$$AE : DH$$
by a ratio of defect which is the square of the former. But by number VIII
$$HI : EM = AE^{3/2} : DH^{3/2}.$$
Therefore let the ratio which is smaller by the total square of the ratio of defect be divided into the ratio of the $3/2$th powers; that is,
$$HI : EM \text{ comp. } GK : HI \text{ comp. } EM : FL \quad AE^{1/2} : DH^{1/2}$$
by the excess squared. But
$$HI : EM \text{ comp. } GK : HI \text{ comp. } EM : FL = GK : FL.$$
Therefore
$$GK : FL \quad AE^{1/2} : DH^{1/2}$$
by the excess squared. But
$$AE : DH = AE : BF \text{ comp. } BF : CG \text{ comp. } CG : DH.$$
And
$$CG : DH \text{ comp. } AE : BF \quad AE^{1/2} : DH^{1/2}$$
by the simple defect. Therefore
$$BF : CG \quad AE^{1/2} : DH^{1/2}$$
by the simple excess. But
$$GK : FL \quad AE^{1/2} : DH^{1/2}$$
but by the excess squared. But the excess squared is greater than the simple excess. Therefore the ratio of the movements GK to FL is greater than the ratio of the corresponding intervals BF to CG.

In fully the same way, it is demonstrated even contrariwise that if the planets approach one another in G and F beyond the mean distances in H and E, in such fashion that the ratio of the mean distances $DH : AE$ becomes less than $DH^{1/2} : AE^{1/2}$, then the ratio of the movements $GK : FL$ becomes less than the ratio of the corresponding intervals $BF : CG$. For you need to do nothing more than to change the words *greater* to *less*, > to <, *excess* to *defect*, and conversely.

In suitable numbers, because the square root of $4/9$ is $2/3$; and $5/8$ is even greater than $2/3$ by the ratio of excess $15/16$; and the square of the ratio 8 : 9 [282] is the ratio 1600 : 2025, *i.e.*, 64 : 81; and the square of the ratio 4 : 5 is the ratio 3456 : 5400, *i.e.*,

16 : 25; and finally the $^3/_2$th power of the ratio 4 : 9 is the ratio 1600 : 5400, *i.e.*, 8 : 27: therefore too the ratio 2025 : 3456, *i.e.*, 75 : 128, is even greater than 5 : 8, *i.e.*, 75 : 120, by the same ratio of excess (*i.e.*, 120 : 128), 15 : 16; whence 2025 : 3456, the ratio of the converging movements, exceeds 5 : 8, the inverse ratio of the corresponding intervals, by as much as 5 : 8 exceeds 2 : 3, the square root of the ratio of the spheres. Or, what amounts to the same thing, the ratio of the two converging intervals is a mean between the ratio of the square roots of the spheres and the inverse ratio of the corresponding movements.

Moreover, from this you may understand that the ratio of the diverging movements is much greater than the ratio of the $^3/_2$th powers of the spheres, since the ratio of the $^3/_2$th powers is compounded with the squares of the ratio of the aphelial interval to the mean interval, and that of the mean to the perihelial.

4. IN WHAT THINGS HAVING TO DO WITH THE PLANETARY MOVEMENTS HAVE THE HARMONIC CONSONANCES BEEN EXPRESSED BY THE CREATOR, AND IN WHAT WAY?

Accordingly, if the image of the retrogradation and stations is taken away and the proper movements of the planets in their real eccentric orbits are winnowed out, the following distinct things still remain in the planets: 1) The distances from the sun. 2) The periodic times. 3) The diurnal eccentric arcs. 4) The diurnal delays in those arcs. 5) The angles at the sun, and the diurnal area apparent to those as it were gazing from the sun. And again, all of these things, with the exception of the periodic times, are variable in the total circuit, most variable at the mean longitudes, but least at the extremes, when, turning away from one extreme longitude, they begin to return to the opposite. Hence when the planet is lowest and nearest to the sun and thereby delays the least in one degree of its eccentric, and conversely in one day traverses the greatest diurnal arc of its eccentric and appears fastest from the sun: then its movement remains for some time in this strength without perceptible variation, until, after passing the perihelion, the planet gradually begins to depart farther from the sun in a straight line; at that same time it delays longer in the degrees of its eccentric circle; or, if you consider the movement of one day, on the following day it goes forward less and appears even more slow from the sun until it has drawn close to the highest apsis and made its distance from the sun very great: for then longest of all does it delay in one degree of its eccentric; or on the contrary in one day it traverses its least arc and makes a much smaller apparent movement and the least of its total circuit.

Finally, all these things may be considered either as they exist in any one planet at different times or as they exist in different planets: whence, by the assumption of an infinite amount of time, all the affects of the circuit of one planet can concur in the same moment of time with all the affects of the circuit of another planet and be compared, and then the total eccentrics, as compared with one another, have the same ratio as their semidiameters or mean intervals; but the arcs of two eccentrics, which are similar or designated by the same number (of degrees), nevertheless have their true lengths unequal in the ratio of their eccentrics. For example, one degree in the sphere of Saturn is approximately twice as long as one degree in the sphere of Jupiter. And conversely, the diurnal arcs of the eccentrics, as expressed in astronomical terms, do not exhibit the ratio of the true journeys which the globes complete in one day [283] through the ether, because the single units in the wider circle of the upper planet denote a quarter part of the journey, but in the narrower circle of the lower planet a smaller part.

Therefore let us take the second of the things which we have posited, namely, the periodic times of the planets, which comprehend the sums made up of all the delays—long, middling, short—in all the degrees of the total circuit. And we found that from antiquity down to us, the planets complete their periodic returns around the sun, as follows in the table:

	Days	Minutes of a day	Therefore the mean diurnal moments		
			Min.	Sec.	Thirds
Saturn	10,759	12	2	0	27
Jupiter	4,332	37	4	59	8
Mars	686	59	31	26	31
Earth with Moon	365	15	59	8	11
Venus	224	42	96	7	39
Mercury	87	58	245	32	25

Accordingly, in these periodic times there are no harmonic ratios, as is easily apparent, if the greater periods are continuously halved, and the smaller are continuously doubled, so that, by neglecting the intervals of an octave, we can investigate the intervals which exist within one octave.

	Saturn	Jupiter	Mars	Earth	Venus	Mercury	
	$10,759^D12'$						
	$5,379^D36'$	$4,332^D37'$				$87^D58'$	
Halves	$2,689^D48'$	$2,166^D19'$			$224^D42'$	$175^D56'$	Doubles
	$1,344^D54'$	$1,083^D10'$	$686^D59'$	$365^D15'$	$449^D24'$	$351^D52'$	
	$672^D27'$	$541^D35'$					

All the last numbers, as you see, are counter to harmonic ratios and seem, as it were, irrational. For let 687, the number of days of Mars, receive as its measure 120, which is the number of the division of the chord: according to this measure Saturn will have 117 for one sixteenth of its period, Jupiter less than 95 for one eighth of its period, the Earth less than 64, Venus more than 78 for twice its period, Mercury more than 61 for four times its period. These numbers do not make any harmonic ratio with 120, but their neighbouring numbers—60, 75, 80, and 96—do. And so, whereof Saturn has 120, Jupiter has approximately 97, the Earth more than 65, Venus more than 80, and Mercury less than 63. And whereof Jupiter has 120, the Earth has less than 81, Venus less than 100, Mercury less than 78. Likewise, whereof Venus has 120, the Earth has less than 98, Mercury more than 94. Finally, whereof the Earth has 120, Mercury has less than 116. But if the free choice of ratios had been effective here, consonances which are altogether perfect but not augmented or diminished would have been taken. Accordingly we find that God the Creator did not wish to introduce harmonic ratios between the sums of the delays added together to form the periodic times.

[284] And although it is a very probable conjecture (as relying on geometrical demonstrations and the doctrine concerning the causes of the planetary movements given in the *Commentaries on Mars*) that the bulks of the planetary bodies are in the ratio of the periodic times, so that the globe of Saturn is about thirty times greater than the globe of the Earth, Jupiter twelve times, Mars less than two, the Earth one and a half times greater than the globe of Venus and four times greater than the globe of Mercury: not therefore will even these ratios of bodies be harmonic.

But since God has established nothing without geometrical beauty, which was not bound by some other prior law of necessity, we easily infer that the periodic times have got their due lengths, and thereby the mobile bodies too have got their bulks, from something which is prior in the archetype, in order to express which thing these bulks and periods have been fashioned to this measure, as they seem disproportionate. But I have said that the periods are added up from the longest, the middling, and the slowest delays: accordingly geometrical fitnesses must be found either in these delays or in anything which may be prior to them in the mind of the Artisan. But the ratios of the delays are bound up with the ratios of the diurnal arcs, because the arcs have the inverse ratio of the delays. Again, we have said that the ratios of the delays and intervals of any one planet are the same. Then, as regards the single planets, there will be one and the same consideration of the following three: the arcs, the delays in equal arcs, and the distance of the arcs from the sun or the intervals. And because all these things are variable in the planets, there can be no doubt but that, if these things were allotted any geometrical beauty, then, by the sure design of the highest Artisan, they would have been received that at their extremes, at the aphelial and perihelial intervals, not at the mean intervals lying in between. For, given the ratios of the extreme intervals, there is no need of a plan to fit the intermediate ratios to a definite number. For they follow of themselves, by the necessity of planetary movement, from one extreme through all the intermediates to the other extreme.

Therefore the intervals are as follows, according to the very accurate observations of Tycho Brahe, by the method given in the *Commentaries on Mars* and investigated in very persevering study for seventeen years.

INTERVALS COMPARED WITH HARMONIC RATIOS[1]

Of Two Planets *Converging Diverging*		Of Single Planets
$^a/_{d}=^2/_1$, $^b/_{c}=^5/_3$	Saturn's aphelion 10,052. a. perihelion 8,968. b.	More than a minor whole tone $^{10,000}/_{9,000}$ Less than a major whole tone $^{10,000}/_{8,935}$
$^c/_{f}=^4/_1$, $^d/_{e}=^3/_1$	Jupiter's aphelion 5,451. c. perihelion 4,949. d.	No concordant ratio but approximately 11 : 10, a discordant or diminished 6 : 5.
$^e/_{h}=^5/_3$, $^f/_{g}=^{17}/_{20}$	Mar's aphelion 1,665. e. perihelion 1,382. f.	Here 1662 : 1385 would be the consonance 6 : 5, and 1665 : 1332 would be 5 : 4
$\dfrac{g}{k}=\dfrac{2}{1\frac{1}{2}}$ viz. $\dfrac{1000}{710}$, $\dfrac{h}{i}=\dfrac{27}{20}$	Earth's aphelion 1,018. g. perihelion 982. h.	Here 1025 : 984 would be the diesis 24 : 25. Therefore it does not have the diesis.
$^i/_{m}=^{12}/_5$, $^k/_{i}=^{243}/_{160}$	Venus' aphelion 729. i. perihelion 719. k.	Less than a sesquicomma. More than one third of a diesis.
	Mercury's aphelion 470. l. perihelion 307. m.	243 : 160, greater than a perfect fifth but less than a harmonic 8 : 5

[285] Therefore the extreme intervals of no one planet come near consonances except those of Mars and Mercury.

But if you compare the extreme intervals of different planets with one another, some harmonic light begins to shine. For the extreme diverging intervals of Saturn and Jupiter make slightly more than the octave; and the converging, a mean between the major and minor sixths. So the diverging extremes of Jupiter and Mars embrace approximately the double octave; and the converging, approximately the fifth and the

1. General Note: Throughout this text Kepler's *concinna* and *inconcinna* are translated as "concordant" and "discordant." *Concinna* is usually used by Kepler of all intervals whose ratios occur within the "natural system" or the just intonation of the scale. *Inconcinna* refers to all ratios that lie outside of this system of tuning. "Consonant" (*consonans*) and "dissonant" (*dissonans*) refer to qualities which can be applied to intervals within the musical system, in other words to "concords." "Harmony" (*harmonia*) is used sometimes in the sense of "concordance" and sometimes in the sense of "consonance."

Genus durum and *genus molle* are translated either as "major mode" and "minor mode," or as "major scale" and "minor scale," or as "major kind" and "minor kind" (of consonances). The use of *modus*, to refer to the ecclesiastical modes, occurs only in Chapter 6.

As our present musical terms do not apply strictly to the music of the sixteenth and seventeenth centuries, a brief explanation of terms here may be useful. This material is taken from Kepler's *Harmonies of the World*, Book III.

An octave system in the minor scale (*Systema octavae in cantu molli*)

| *Ratios of string lengths:* | g
72 : | f
81 : | e
90 : | d
96 : | c
108 : | b
120 : | A
128 : | G
144 |

octave. But the diverging extremes of the Earth and Mars embrace somewhat more than the major sixth; the converging, an augmented fourth. In the next couple, the Earth and Venus, there is again the same augmented fourth between the converging extremes; but we lack any harmonic ratio between the diverging extremes: for it is less than the semi-octave (so to speak), *i.e.*, less than the square root of the ratio 2 : 1. Finally, between the diverging extremes of Venus and Mercury there is a ratio slightly less than the octave compounded with the minor third; between the converging there is a slightly augmented fifth.

Continued from page 21

In the major scale (*In cantu duro*)

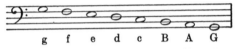

g   f   e   d   c   B   A   G

Ratios of string lengths: 360 : 405 :   432 :   480 :   540 :   576 :   640 :   720
As in all music, these scales can be repeated at one or more octaves above. The ratios would then all be halved, *i.e.*,

g′   f′   e′   d′   c′   b   a   g   f

180 :   202½ : 216 :   240 :   270 :   288 :   320 :   360 :   405 etc.

Various intervals which Kepler considers are:

Ratio	Name	Description
80 : 81	*comma* (of Didymus), difference between major and minor whole tones (8/9 ÷ 9/10)	
24 : 25	*diesis* [difference between e – e flat or B – b flat or between a semitone and a minor whole tone (15/16 ÷ 9/10)]	
128 : 135	*lemma* [difference between a semitone and a major whole tone (15/16 ÷ 8/9)]	
243 : 256	*Plato's lemma* (not found in this system but in the Pythagorean tuning)	
15 : 16	*semitone*	minor mode between e flat – d, b flat – A major mode between e – d, B – A
9 : 10	*minor whole tone*	minor mode f – e flat, c – b flat major mode e – d, B – A
8 : 9	*major whole tone*	minor mode: g – f, d – c, A – G major mode: g – f, d – c, A – G
27 : 32	*sub-minor third*	(major and minor modes: f – d, c – A)
5 : 6	*minor third*	minor mode: e – flat – c, b flat – G major mode: g – e, d – B
4 : 5	*major third*	minor mode: g – e – flat, d – b – flat major mode: e – c, B – G
64 : 81	*ditone* (Pythagorean third) (major and minor modes: a – f)	
243 : 320	*lesser imperfect fourth* (inversion of "greater imperfect fifth") see below	
3 : 4	*perfect fourth*	minor mode: g – d, f – c, e flat – b flat, d – A, c – G major mode: g – d, f – c, e – B, , d – A, c – G
20 : 27	*greater imperfect fourth*	minor mode: b′ flat – f major mode: a – e
32 : 45	*augmented fourth*	minor mode: a – e flat major mode: b – f
45 : 64	*diminished fifth*	minor mode: e – flat – A major mode: f – B
27 : 40	*lesser imperfect fifth*	minor mode: f – b flat major mode: e – A
2 : 3	*perfect fifth*	minor mode: g – c, d – G major mode: g – c, d – G
160 : 243	*greater imperfect fifth* (compound of ditone and minor third 64/81 x 5/6)	
81 : 128	*imperfect minor sixth* (minor and major modes: f – A)	
5 : 8	*minor sixth*	minor mode: e flat – G, b^{1b+} – d major mode: g – B, c′ – e
3 : 5	*major sixth*	minor mode: g – B flat, c′ – e flat major mode: e – G, b – d
64 : 27	*greater major sixth*	minor mode: d′ – f, a – c major mode: d′ – f, a – c
1 : 2	*octave* (g – G, a – A, b – B, b flat – b flat)	

All these are simple intervals. When one or more octaves are added to any simple intervals the resultant interval is a "compound" interval.
1 : 3 equals 1/2 x 2/3—an octave and a perfect fifth
3 : 32 equals (1/2)³ x 3/4—three octaves and a perfect fourth
1 : 20 equals (1/2)⁴ x 16/20—four octaves and a major third

Accordingly, although one interval was somewhat removed from harmonic ratios, this success was an invitation to advance further. Now my reasonings were as follows: First, in so far as these intervals are lengths without movement, they are not fittingly examined for harmonic ratios, because movement is more properly the subject of consonances, by reason of speed and slowness. Second, inasmuch as these same intervals are the diameters of the spheres, it is believable that the ratio of the five regular solids applied proportionally is more dominant in them, because the ratio of the geometrical solid bodies to the celestial spheres (which are everywhere either encompassed by celestial matter, as the ancients hold, or to be encompassed successively by the accumulation of many revolutions) is the same as the ratio of the plane figures which may be inscribed in a circle (these figures engender the consonances) to the celestial circles of movements and the other regions wherein the movements take place. Therefore, if we are looking for consonances, we should look for them not in these intervals in so far as they are the semidiameters of spheres but in them in so far as they are the measures of the movements, *i.e.*, in the movements themselves, rather. Absolutely no other than the mean intervals can be taken as the semidiameters of the spheres; but we are here dealing with the extreme intervals. Accordingly, we are not dealing with the intervals in respect to their spheres but in respect to their movements.

Accordingly, although for these reasons I had passed on to the comparison of the extreme movements, at first the ratios of the movements remained the same in magnitude as those which were previously the ratios of the intervals, only inverted. Wherefore too, certain ratios, which are discordant and foreign to harmonies, as before, have been found between the movements. But once again I judged that this happened to me deservedly, because I compared with one another eccentric arcs which

Continued from page 22

Concords: All intervals from diesis downward on above list.
Consonances: Minor and major thirds and sixths, perfect fourth, fifth, and octave.
"Adulterine" consonances: sub-minor third, ditone, lesser imperfect fourth and fifth, greater imperfect fourth and fifth, imperfect minor sixth, greater major sixth.
Dissonances: All other intervals.
Throughout this work Kepler, after the fashion of the theorists of his time, uses the ratios of string lengths rather than the ratios of vibrations as is usually done today. String lengths are, of course, inversely proportionate to the vibrations. That is, string lengths 4 : 5 are expressed in vibrations as 5 : 4. This accounts for the descending order of the scale, which follows the increasing numerical order. It is an interesting fact that Kepler's minor and major scales are inversions of each other and hence, when expressed in ratios of vibrations, are in the opposite order from those in ratios of string lengths:

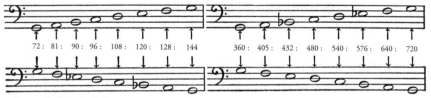

72 : 81 : 90 : 96 : 108 : 120 : 128 : 144   360 : 405 : 432 : 480 : 540 : 576 : 640 : 720

Notes resulting from ratios of vibrations
Notes resulting from ratios of string lengths

An arbitrary pitch G is chosen to situate these ratios. This g or "gamma" was usually the lowest tone of the sixteenth-century musical gamut.
Elliott Carter, Jr.

are not expressed and numbered by a measure of the same magnitude but are numbered in degrees and minutes which are of diverse magnitude in diverse planets, nor do they from our place give the appearance of being as great as the number of each says, except only at the centre of the eccentric of each planet, which centre rests upon no body; and hence it is also unbelievable that there is any sense or natural instinct in that place in the world which is capable of perceiving this; or, rather, it was impossible, if I was comparing the eccentric arcs of different planets with respect to their appearance at their centres, which are different for different planets. But if diverse apparent magnitudes are compared with one another, they ought to be apparent in one place in the world in such a way that that which possesses the faculty of comparing them may be present in that place from which they are all apparent. Accordingly, I judged that the appearance of these eccentric arcs should be removed from the mind or else should be formed differently. But if I removed the appearance and applied my mind to the diurnal journeys of the planets, I saw that I had to employ the rule which I gave in Article IX of the preceding chapter. [286] Accordingly if the diurnal arcs of the eccentric are multiplied by the mean intervals of the spheres, the following journeys are produced:

		Diurnal movements	Mean intervals	Diurnal journeys
Saturn	at aphelion	1'53"	9510	1065
	at perihelion	2'7"		1208
Jupiter	at aphelion	4'44"	5200	1477
	at perihelion	5'15"		1638
Mars	at aphelion	28'44"	1524	2627
	at perihelion	34'34"		3161
Earth	at aphelion	58'6"	1000	3486
	at perihelion	60'13"		3613
Venus	at aphelion	95'29"	724	4149
	at perihelion	96'50"		4207
Mercury	at aphelion	201'0"	388	4680
	at perihelion	307'3"		7148

Thus Saturn traverses barely one seventh of the journey of Mercury; and hence, as Aristotle judged consonant with reason in Book II of *On the Heavens*, the planet which is nearer the sun always traverses a greater space than the planet which is farther away—as cannot hold in the ancient astronomy.

And indeed, if we weigh the thing fairly carefully, it will appear to be not very probable that the most wise Creator should have established harmonies between the planetary journeys in especial. For if the ratios of the journeys are harmonic, all the other affects which the planets have will be necessitated and bound up with the journeys, so that there is no room elsewhere for establishing harmonies. But whose good will it be to have harmonies between the journeys, or who will perceive these harmonies? For there are two things which disclose to us harmonies in natural things: either light or sound: light apprehended through the eyes or hidden senses proportioned to the eyes, and sound through the ears. The mind seizes upon these forms and, whether by instinct (on which Book IV speaks profusely) or by astronomical or harmonic ratiocination, discerns the concordant from the discordant. Now there are no sounds in the heavens, nor is the movement so turbulent that any noise is made by the rubbing against the ether. Light remains. If light has to teach these things about the planetary journeys, it will teach either the eyes or a sensorium analogous to the eyes and situated in a definite place; and it seems that sense-perception must be present there in order that light of itself may immediately teach. Therefore there will be sense-perception in the total world, namely in order that the movements of all the planets may be presented to sense-perceptions at the same time. For that former route—from observations through the longest detours of geometry and arithmetic, through the ratios of the spheres and the other things which must be learned first, down to the journeys which have been exhibited—is too long for any natural instinct, for the sake of moving which it seems reasonable that the harmonies have been introduced.

Therefore with everything reduced to one view, I concluded rightly [287] that the true journeys of the planets through the ether should be dismissed, and that we should turn our eyes to the apparent diurnal arcs, according as they are all apparent from one definite and marked place in the world—namely, from the solar body itself, the source of movement of all the planets; and we must see, not how far away from the sun any one of the planets is, nor how much space it traverses in one day (for that is something for ratiocination and astronomy, not for instinct), but how great an angle the diurnal movement of each planet subtends in the solar body, or how great an arc it seems to traverse in one common circle described around the sun, such as the ecliptic, in order that these appearances, which were conveyed to the solar body by virtue of light, may be able to flow, together with the light, in a straight line into creatures, which are partakers of this instinct, as in Book IV we said the figure of the heavens flowed into the foetus by virtue of the rays.

Therefore, if you remove from the proper planetary movement the parallaxes of the annual orbit, which gives them the mere appearances of stations and retrogradations, Tycho's astronomy teaches that the diurnal movements of the planets in their orbits (which are apparent as it were to spectator at the sun) are as shown in the table on the next page.

Note that the great eccentricity of Mercury makes the ratio of the movements differ somewhat from the ratio of the square of the distances. For if you make the square of the ratio of 100, the mean distance, to 121, the aphelial distance, be the ratio of the aphelial movement to the mean movement of 245'32", then an aphelial movement of 167 will be produced; and if the square of the ratio of 100 to 79, the perihelial distance, be the ratio of the perihelial to the same mean movement, then the perihelial movement will become 393; and both cases are greater than I have here laid down, because the mean movement at the mean anomaly, viewed very obliquely, does not appear as great, *viz.*, not as great as 245'32", but about 5' less. Therefore, too, lesser aphelial and perihelial movements will be elicited. But the aphelial (appears) lesser and the perihelial greater, on account of theorem 8, Euclid's *Optics*, as I remarked in the preceding Chapter, Article VI.

Harmonies Between Two Planets		Apparent Diurnal Movements			Harmonies Between the Movements of Single Planets
Diverging	*Converging*				
		Saturn	at aphelion	1'46" a.	1 : 48" : 2'15" = 4 : 5,
			at perihelion	2'15" b.	major third
$^a/_{d}=^1/_3,$	$^b/_{c}=^1/_2$				
		Jupiter	at aphelion	4'30" c.	4'35" : 5'30" = 5 : 6,
			at perihelion	5'30" d.	minor third
$^c/_{f}=^1/_8,$	$^d/_{e}=^5/_{24}$				
		Mars	at aphelion	26'14" e.	25'21" : 38'1" = 2 : 3,
			at perihelion	38'1" f.	the fifth
$^e/_{h}=^5/_{12},$	$^f/_{g}=^2/_3$				
		Earth	at aphelion	57'3" g.	57'28" : 61'18" =
			at perihelion	61'18" h.	15 : 16, semitone
$^g/_{k}=^3/_5,$	$^h/_{i}=^5/_8$				
		Venus	at aphelion	94'50" i.	94'50" : 98'47" =
			at perihelion	97'37" k.	24 : 25, diesis
$^i/_{m}=^1/_4,$	$^k/_{l}=^3/_5$				
		Mercury	at aphelion	164'0" l.	164'0" : 394'0" = 5 : 12,
			at perihelion	384'0" m.	octave and minor third

Accordingly, I could mentally presume, even from the ratios of the diurnal eccentric arcs given above, that there were harmonies and concordant intervals between these extreme apparent movements of the single planets, since I saw that everywhere there the square roots of harmonic ratios were dominant, but knew that the ratio of the apparent movements was the square of the ratio of the eccentric movements. But it is possible by experience itself, or without any ratiocination to prove what is affirmed, as you see [288] in the preceding table. The ratios of the apparent movements of the single planets approach very close to harmonies, in such fashion that Saturn and Jupiter embrace slightly more than the major and minor thirds, Saturn with a ratio of excess of 53 : 54, and Jupiter with one of 54 : 55 or less, namely approximately a sesquicomma; the Earth, slightly more (namely 137 : 138, or barely a semicomma) than a semitone; Mars somewhat less (namely 29 : 30, which approaches 34 : 35 or 35 : 36) than a fifth; Mercury exceeds the octave by a minor third rather than a whole tone, viz., it is about 38 : 39 (which is about two commas, viz., 34 : 35 or 35 : 36) less than a whole tone. Venus alone falls short of any of the concords the diesis; for its ratio is between two and three commas, and it exceeds two thirds of a diesis, and is about 34 : 35 or 35 : 36, a diesis diminished by a comma.

The moon, too, comes into this consideration. For we find that its hourly apogeal movement in the quadratures, viz., the slowest of all its movements, to be 26'26"; its perigeal movement in the syzygies, viz., the fastest of all, 35'12", in which way the perfect fourth is formed very precisely. For one third of 26'26" is 8'49", the quadruple of which is 35'16". And note that the consonance of the perfect fourth is found nowhere else between the apparent movements; note also the analogy between the fourth in consonances and the quarter in the phases. And so the above things are found in the movements of the single planets.

But in the extreme movements of two planets compared with one another, the radiant sun of celestial harmonies immediately shines at first glance, whether you compare the diverging extreme movements or the converging. For the ratio between the diverging movements of Saturn and Jupiter is exactly the duple or octave; that between the diverging, slightly more than triple or the octave and the fifth. For one third of 5'30" is 1'50", although Saturn has 1'46" instead of that. Accordingly, the planetary movements will differ from a consonance by a diesis more or less, viz., 26 : 27 or 27 : 28; and with less than one second acceding at Saturn's aphelion, the excess will be 34 : 35, as great as the ratio of the extreme movements of Venus. The diverging and converging movements of Jupiter and Mars are under the sway of the triple octave and the double octave and a third, but not perfectly. For one eighth of 38'1" is 4'45", although Jupiter has 4'30"; and between these numbers there is still a difference of 18 : 19, which is a mean between the semitone of 15 : 16 and the diesis of 24 : 25, namely, approximately a perfect lemma of 128 : 135.[1] Thus one fifth of 26'14" is 5'15", although Jupiter has 5'30"; accordingly in this case the quintuple ratio is diminished in the ratio of 21 : 22, the augment in the case of the other ratio, viz., approximately a diesis of 24 : 25.

1. cf. Footnote to *Intervals Compared with Harmonic Ratios*, p. 186.

The consonance 5 : 24 comes nearer, which compounds a minor instead of a major third with the double octave. For one fifth of 5'30" is 1'6", which if multiplied by 24 makes 26'24", does not differ by more than a semicomma. Mars and the Earth have been allotted the least ratio, exactly the sesquialteral or perfect fifth: for one third of 57'3" is 19'1", the double of which is 38'2", which is Mars' very number, *viz.*, 38'11". They have also been allotted the greater ratio of 5 : 12, the octave and minor third, but more imperfectly. For one twelfth of 61'18" is 5'6½," which if multiplied by 5 gives 25'33", although instead of that Mars has 26'14". Accordingly, there is a deficiency of a diminished diesis approximately, *viz.*, 35 : 36. But the Earth and Venus together have been allotted 3 : 5 as their greatest consonance and 5 : 8 as their least, the major and minor sixths, but again not perfectly. For one fifth of 97'37", which if multiplied by 3 gives 58'33", which is greater than the movement of the Earth in the ratio 34 : 35, which is approximately 35 : 36: by so much do the planetary ratios differ from the harmonic. Thus one eighth of 94'50" is 11'51"+, five times which is 59'16", which is approximately equal to the mean movement of the Earth. Wherefore here the planetary ratio is less than the harmonic [289] in the ratio of 29 : 30 or 30 : 31, which is again approximately 35 : 36, the diminished diesis; and thereby this least ratio of these planets approaches the consonance of the perfect fifth. For one third of 94'50" is 31'37", the double of which is 63'14", of which the 61'18" of the perihelial movement of the Earth falls short in the ratio of 31 : 32, so that the planetary ratio is exactly a mean between the neighbouring harmonic ratios. Finally, Venus and Mercury have been allotted the double octave as their greatest ratio and the major sixth as their least, but not absolute-perfectly. For one fourth of 384' is 96'0", although Venus has 94'50". Therefore the quadruple adds approximately one comma. Thus one fifth of 164' is 32'48", which if multiplied by 3 gives 98'24", although Venus has 97'37". Therefore the planetary ratio is diminished by about two thirds of a comma, *i.e.*, 126 : 127.

Accordingly the above consonances have been ascribed to the planets; nor is there any ratio from among the principal comparisons (*viz.*, of the converging and diverging extreme movements) which does not approach so nearly to some consonance that, if strings were tuned in that ratio, the ears would not easily discern their imperfection—with the exception of that one excess between Jupiter and Mars.

Moreover, it follows that we shall not stray far away from consonances if we compare the movements of the same field. For if Saturn's 4 : 5 comp. 53 : 54 are compounded with the intermediate 1 : 2, the product is 2 : 5 comp. 53 : 54, which exists between the aphelial movements of Saturn and Jupiter. Compound with that Jupiter's 5 : 6 comp. 54 : 55, and the product is 5 : 12 comp. 54 : 55, which exist between the perihelial movements of Saturn and Jupiter. Thus compound Jupiter's 5 : 6 comp.

54 : 55 with the intermediate ensuing ratio of 5 : 24 comp. 158 : 157, the product will be 1 : 6 comp. 36 : 35 between the aphelial movements. Compound the same 5 : 24 comp. 158 : 157 with Mars' 2 : 3 comp. 30 : 29, and the product will be 5 : 36 comp. 25 : 24 approximately, *i.e.*, 125 : 864 or about 1 : 7, between the perihelial movements. This ratio is still alone discordant. With 2 : 3 the third ratio among the intermediates, compound Mars' 2 : 3 less 29 : 30; the result will be 4 : 9 comp. 30 : 29, *i.e.*, 40 : 87, another discord between the aphelial movements. If instead of Mars' you compound the Earth's 15 : 16 comp. 137 : 138, you will make 5 : 8 comp. 137 : 138 between the perihelial movements. And if with the fourth of the intermediates, 5 : 8 comp. 31 : 30, or 2 : 3 comp. 31 : 32, you compound the Earth's 15 : 16 comp. 137 : 138, the product will be approximately 3 : 5 between the aphelial movements of the Earth and Venus. For one fifth of 94'50" is 18'58", the triple of which is 56'54", although the Earth has 57'3". If you compound Venus' 34 : 35 with the same ratio, the result will be 5 : 8 between the perihelial movements. For one eighth of 97'37" is 12'12"+ which if multiplied by 5 gives 61'1", although the Earth has 61'18". Finally, if with the last of the intermediate ratios, 3 : 5 comp. 126 : 127 you compound Venus' 34 : 35, the result is 3 : 5 comp. 24 : 25, and the interval, compounded of both, between the aphelial movements, is dissonant. But if you compound Mercury's 5 : 12 comp. 38 : 39, the double octave or 1 : 4 will be diminished by approximately a whole diesis, in proportion to the perihelial movements.

Accordingly, perfect consonances are found: between the converging movements of Saturn and Jupiter, the octave; between the converging movements of Jupiter and Mars, the octave and minor third approximately; between the converging movements of Mars and the Earth, the fifth; between their perihelial, the minor sixth; between the extreme converging movements of Venus and Mercury, the major sixth; between the diverging or even between the perihelial, the double octave: whence without any loss to an astronomy which has been built, most subtly of all, upon Brahe's observations, it seems that the residual very slight discrepancy can be discounted, especially in the movements of Venus and Mercury.

But you will note that where there is no perfect major consonance, as between Jupiter and Mars, there alone have I found the placing of the solid figure to be approximately perfect, since the perihelial distance of Jupiter is approximately three times the aphelial distance of Mars, in such fashion that this pair of planets strives after the perfect consonance in the intervals which it does not have in the movements.

[290] You will note, furthermore, that the major planetary ratio of Saturn and Jupiter exceeds the harmonic, *viz.*, the triple, by approximately the same quantity as belongs to Venus; and the common major ratio of the converging and diverging movements of Mars and the Earth are diminished by approximately the same. You will note

thirdly that, roughly speaking, in the upper planets the consonances are established between the converging movements, but in the lower planets, between movements in the same field. And note fourthly that between the aphelial movements of Saturn and the Earth there are approximately five octaves; for one thirty-second of 57'3" is 1'47", although the aphelial movement of Saturn is 1'46".

Furthermore, a great distinction exists between the consonances of the single planets which have been unfolded and the consonances of the planets in pairs. For the former cannot exist at the same moment of time, while the latter absolutely can; because the same planet, moving at its aphelion, cannot be at the same time at the opposite perihelion too, but of two planets one can be at its aphelion and the other at its perihelion at the same moment of time. And so the ratio of plain-song or monody, which we call choral music and which alone was known to the ancients,[1] to polyphony—called "figured song,";[2] the invention of the latest generations—is the same as the ratio of the consonances which the single planets designate to the consonances of the planets taken together. And so, further on, in Chapters 5 and 6, the single planets will be compared to the choral music of the ancients and its properties will be exhibited in the planetary movements. But in the following chapters, the planets taken together and the figured modern music will be shown to do similar things.

5. IN THE RATIOS OF THE PLANETARY MOVEMENTS WHICH ARE APPARENT AS IT WERE TO SPECTATORS AT THE SUN, HAVE BEEN EXPRESSED THE PITCHES OF THE SYSTEM, OR NOTES OF THE MUSICAL SCALE, AND THE MODES OF SONG (GENERA CANTUS), THE MAJOR AND THE MINOR[3]

Therefore by now I have proved by means of numbers gotten on one side from astronomy and on the other side from harmonies that, taken in every which way, harmonic ratios hold between these twelve termini or movements of the six planets revolving around the sun or that they approximate such ratios within an imperceptible part of least concord. But just as in Book III in the first chapter, we first built up the single harmonic consonances separately, and then we joined together all the consonances—as many as there were—in one common system or musical scale, or, rather, in one octave of them which embraces the rest in power, and by means of them we separated the others into their degrees or pitches (*loca*) and we did this in such a way that there would be a scale; so now also, after the discovery of the consonances (*harmoniis*) which

---

1. The choral music of the Greeks was monolinear, everyone singing the same melody together—E. C., Jr.
2. In plain-song all the time values of the notes were approximately equal, while in "figured song" time values of different lengths were indicated by the notes, which gave composers an opportunity both to regulate the way different contrapuntal parts joined together and to produce many expressive effects. Practically all melodies since this time are in "figured song" style.—E. C., Jr.
3. See note to *Intervals Compared with Harmonic Ratios.*

God Himself has embodied in the world, we must consequently see whether those single consonances stand so separate that they have no kinship with the rest, or whether all are in concord with one another. Notwithstanding it is easy to conclude, without any further inquiry, that those consonances were fitted together by the highest prudence in such fashion that they move one another about within one frame, so to speak, and do not jolt one another out of it; since indeed we see that in such a manifold comparison of the same terms there is no place where consonances do not occur. For unless in one scale all the consonances were fitted to all, it could easily have come about (and it has come about wherever necessity thus urges it) that many dissonances should exist. For example, if someone had set up a major sixth between the first and the second term, and likewise a major third between the second and the third term, without taking the first into account, then he would admit a dissonance and the discordant interval 12 : 25 between the first and third.

But come now, let us see whether that which we have already inferred by reasoning is really found in this way. [291] But let me premise some cautions, that we may be the less impeded in our progress. First, for the present, we must conceal those augments or diminutions which are less than a semitone; for we shall see later on what causes they have. Second, by continuous doubling or contrary halving of the movements, we shall bring everything within the range of one octave, on account of the sameness of consonance in all the octaves.

Accordingly the numbers wherein all the pitches or clefs (*loca seu claves*) of the octave system are expressed have been set out in a table in Book III, Chapter 7[1], *i.e.*,

1. The table is as follows:

Concordant Intervals	Lengths of Strings	In familiar notes
	1080	High g
Semitone		
	1152	f ♯
Lemma		
	1215	f
Semitone		
	1296	e
Diesis		
	1350	e ♭
Semitone		
	1440	d
Semitone		
	1536	c ♯
Lemma		
	1620	c
Semitone		
	1728	b
Diesis		
	1800	b ♭
Semitone		
	1920	A
Semitone		
	2048	G ♯
Lemma		
	2160	Low G

31

understand these numbers of the length of two strings. As a consequence, the speeds of the movements will be in the inverse ratios.

Now let the planetary movements be compared in terms of parts continuously halved. Therefore

Movement of Mercury	at perihelion,	7th subduple, or $1/128$,	3'0"
	at aphelion,	6th subduple, or $1/64$,	2'34"
Movement of Venus	at perihelion,	5th subduple, or $1/32$,	3'3"
	at aphelion,	5th subduple, or $1/32$,	2'58"
Movement of Earth	at perihelion,	5th subduple, or $1/32$,	1'55"
	at aphelion,	5th subduple, or $1/32$,	1'47"
Movement of Mars	at perihelion,	4th subduple, or $1/16$,	2'23"
	at aphelion,	3rd subduple, or $1/8$,	3'17"
Movement of Jupiter	at perihelion,	subduple, or $1/2$,	2'45"
	at aphelion,	subduple, or $1/2$	2'15"
Movement of Saturn	at perihelion,		2'15"
	at aphelion,		1'46"

Now the aphelial movement of Saturn at its slowest—*i.e.*, the slowest movement—marks *G*, the lowest pitch in the system with the number 1'46". Therefore the aphelial movement of the Earth will mark the same pitch, but five octaves higher, because its number is 1'47", and who wants to quarrel about one second in the aphelial movement of Saturn? But let us take it into account, nevertheless; the difference will not be greater than 106 : 107, which is less than a comma. If you add 27", one quarter of this 1'47", the sum will be 2'14", although the perihelial movement of Saturn has 2'15"; similarly the aphelial movement of Jupiter, but one octave higher. Accordingly, these two movements mark the note *b*, or else are very slightly higher. Take 36", one third of 1'47", and add it to the whole; you will get as a sum 2'23" for the note *c*; and here's the perihelion of Mars of the same magnitude but four octaves higher. To this same 1'47" add also 54", half of it, and the sum will be 2'41" for the note *d*; and here the perihelion of Jupiter is at hand, but one octave higher, for it occupies the nearest number, *viz.*, 2'45". If you add two thirds, *viz.*, 1'11", the sum will be 2'58"; and here's the aphelion of Venus at 2'58". Accordingly, it will mark the pitch or the note *e*, but five octaves higher. And the perihelial movement of Mercury, which is 3'0", does not exceed it by much but is seven octaves higher. Finally, divide the double of 1'47", *viz.*, 3'34", into nine parts and subtract one part of 24" from the whole; 3'10" will be left for the note *f*, which the 3'17" of the aphelial movement of Mars marks approximately but three octaves higher; and this number is slightly greater than the just number and approaches the note *f* sharp. For if one sixteenth of 3'34", *viz.*, $13^{1}/_{2}$", is subtracted from 3'34", then $3'20^{1}/_{2}$" is left, to which 3'17" is very near. And indeed in music *f* sharp is often employed in place of *f*, as we can see everywhere.

Accordingly all the notes of the major scale (*cantus duri*) (except the note *a* which was not marked by harmonic division, in Book III, Chapter 2) are marked by all the extreme movements of the planets, except the perihelial movements of Venus and the Earth [292] and the aphelial movement of Mercury, whose number, 2'34", approaches the note *c* sharp. For subtract from the 2'41" of *d* one sixteenth or 10", and 2'30" remains for the note *c* sharp. Thus only the perihelial movement of Venus and the Earth are missing from this scale, as you may see in the table.

On the other hand, if the beginning of the scale is made at 2'15", the aphelial

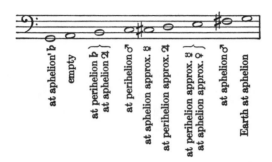

movement of Saturn, and we must express the note *G* in those degrees: then for the note *A* is 2'32", which closely approaches the aphelial movement of Mercury; for the note *b* flat, 2'42", which is approximately the perihelial movement of Jupiter, by the equipollence of octaves; for the note *c*, 3'0", approximately the perihelial movement of Mercury and Venus; for the note *d*, 3'23" and the aphelial movement of Mars is not much graver, *viz.*, 3'17", so that here the number is about as much less than its note as previously the same number was greater than its note; for the note *e* flat, 3'36", which the aphelial movement of the Earth approximates; for the note *e*, 3'50", and the perihelial movement of the Earth is 3'49"; but the aphelial movement of Jupiter again occupies *g*. In this way, all the notes except *f* are

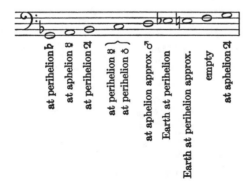

expressed within one octave of the minor scale by most of the aphelial and perihelial movements of the planets, especially by those which were previously omitted, as you see in the table.

Previously, however, *f* sharp was marked and *a* omitted; now *a* is marked, *f* sharp is omitted; for the harmonic division in Chapter 2 also omitted the note *f*.

Accordingly, the musical scale or system of one octave with all its pitches, by means of which natural song[1] is transposed in music, has been expressed in the heavens by a twofold way and in two as it were modes of song. There is this sole difference: in our harmonic sectionings both ways start together from one and the same terminus *G*; but here, in the planetary movements, that which was previously *b* now becomes *G* in the minor mode.

In the celestial movements, as follows:

By harmonic sectionings, as follows:

For as in music 2160 : 1800, or 6 : 5, so in that system which the heavens express, 1728 : 1440, namely, also 6 : 5; and so for most of the remaining, 2160 : 1800, 1620, 1440, 1350, 1080 as 1728 : 1440, 1296, 1152, 1080, 864.

Accordingly you won't wonder any more that a very excellent order of sounds or pitches in a musical system or scale has been set up by men, since you see that they are doing nothing else in this business except to play the apes of God the Creator and to act out, as it were, a certain drama of the ordination of the celestial movements.

But there still remains another way whereby we may understand the twofold musical scale in the heavens, where one and the same system but a twofold tuning (*tensio*) is embraced, one at the aphelial movement of Venus, the other at the perihelial, because the variety of movements of this planet is of the least magnitude, as being such as is comprehended within the magnitude of the diesis, the least concord. And the aphelial tuning (*tensio*), as above, has been given to the aphelial movements of Saturn, the Earth, Venus, and (relatively speaking) Jupiter, in *G*, *e*, *b*, but to the perihelial movements of Mars and (relatively speaking) Saturn and, as is apparent at first glance, to

---

1. Natural song: music in the basic major or minor system without accidentals. E. C., Jr.

those of Mercury, in *c*, *e*, and *b*. On the other hand, the perihelial tuning supplies a pitch even for the aphelial movements of Mars, Mercury, and (relatively speaking) Jupiter, but to the perihelial movements of Jupiter, Venus, and (relatively speaking) Saturn, and to a certain extent to that of the Earth and indubitably to that of Mercury too. For let us suppose that now not the aphelial movement of Venus but the 3'3" of the perihelial gets the pitch of *e*; it is approached very closely by the 3'0" of the perihelial movement of Mercury, through a double octave, at the end of Chapter 4. But if 18" or one tenth of this perihelial movement of Venus is subtracted, 2'45" remains, the perihelion of Jupiter, which occupies the pitch of *d*; and if one fifteenth or 12" is added, the sum will be 3'15", approximately the perihelion of Mars which occupies the pitch of *f*, and thus in *b*, the perihelial movement of Saturn and the aphelial movement of Jupiter have approximately the same tuning. But one eighth, or 23", if multiplied by 5, gives 1'55", which is the perihelial movement of the Earth; and, although it does not square with the foregoing in the same scale, as it does not give the interval 5 : 8 below *e* nor 24 : 25 above *G*, nevertheless if now the perihelial movement of Venus and so too the aphelial movement of Mercury, outside of the order, occupy the pitch *e*-flat instead of *e*, then there the perihelial movement of the Earth will occupy the pitch of *G*, and the aphelial movement of Mercury is in concord, because 1'1", or one third of 3'3", if multiplied by 5, gives 5'5", half of which, or 2'32", approximates the aphelion of Mercury, which in this extraordinary adjustment will occupy the pitch of *c*. Therefore, all these movements are of the same tuning with respect to one another; but the perihelial movement of Venus together with the three (or five) prior movements, *viz.*, in the same harmonic mode, divides the scale differently from the aphelial movement of the same in its tuning, *viz.*, in the major mode (*denere duro*). Moreover, the perihelial movement of Venus, together with the two posterior movements, divides the same scale differently, *viz.*, not into concords but merely into a different order of concords, namely one which belongs to the minor mode (*generis mollis*).

But it is sufficient to have laid before the eyes in this chapter what is the case casually, but it will be disclosed in Chapter 9 by the most lucid demonstrations why each and every one of these things was made in this fashion and what the causes were not merely of harmony but even of the very least discord.

6. IN THE EXTREME PLANETARY MOVEMENTS THE MUSICAL MODES OR TONES HAVE SOMEHOW BEEN EXPRESSED

[294] This follows from the aforesaid and there is no need of many words; for the single planets somehow mark the pitches of the system with their perihelial movement, in so far as it has been appointed to the single planets to traverse a certain fixed interval in the musical scale comprehended by the definite notes of it or the pitches of the

system, and beginning at that note or pitch of each planet which in the preceding chapter fell to the aphelial movement of that planet: *G* to Saturn and the Earth, *b* to Jupiter, which can be transposed higher to *G*, *f*-sharp to Mars, *e* to Venus, *a* to Mercury in the higher octave. See the single movements in the familiar terms of notes. They do not form articulately the intermediate positions, which you here see filled by notes, as they do the extremes, because they struggle from one extreme to the opposite not by leaps and intervals but by a continuum of tunings and actually traverse all the means (which are potentially infinite)—which cannot be expressed by me in any other way than by a continuous series of intermediate notes. Venus remains approximately in unison and does not equal even the least of the concordant intervals in the difference of its tension.

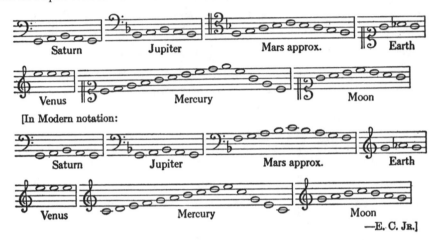

Saturn   Jupiter   Mars approx.   Earth

Venus   Mercury   Moon

[In Modern notation:

Saturn   Jupiter   Mars approx.   Earth

Venus   Mercury   Moon

—E. C. JR.]

But the signature of two accidentals (flats) in a common staff and the formation of the skeletal outline of the octave by the inclusion of a definite concordant interval are a certain first beginning of the distinction of Tones or Modes (*modorum*). Therefore the musical Modes have been distributed among the planets. But I know that for the formation and determination of distinct Modes many things are requisite, which belong to human song, as containing (a) distinct (order of) intervals; and so I have used the word *somehow*.

But the harmonist will be free to choose his opinion as to which Mode each planet expresses as its own, since the extremes have been assigned to it here. From among the familiar Modes, I should give to Saturn the Seventh or Eighth, because if you place its key-note at *G*, the perihelial movement ascends to *b*; to Jupiter, the First or Second Mode, because its aphelial movement has been fitted to *G* and its perihelial movement arrives at *b* flat; to Mars, the Fifth or Sixth Mode, not only because Mars comprehends approximately the perfect fifth, which interval is common to all the Modes, but principally because when it is reduced with the others to a common system, it attains *c* with its perihelial movement and touches *f* with its aphelial, which is the key-note of the Fifth or Sixth Mode or Tone; I should give the Third or Fourth Mode to the Earth, because its movement revolves within a semitone, while the first interval of those Modes is a semitone; but to Mercury will belong indifferently all the Modes or Tones on account of the greatness of its range; to Venus, clearly none on account of the smallness of its range; but on account of the common system the Third and Fourth Mode, because with reference to the other planets it occupies *e*. (The Earth sings MI, FA, MI so that you may infer even from the syllables that in this our domicile MIsery and FAmine obtain.)[1]

7. THE UNIVERSAL CONSONANCES OF ALL SIX PLANETS, LIKE COMMON FOUR-PART COUNTERPOINT, CAN EXIST

[295] But now, Urania, there is need for louder sound while I climb along the harmonic scale of the celestial movements to higher things where the true archetype of the fabric of the world is kept hidden. Follow after, ye modern musicians, and judge the thing according to your arts, which were unknown to antiquity. Nature, which is never not lavish of herself, after a lying-in of two thousand years, has finally brought you forth in these last generations, the first true images of the universe. By means of your concords of various voices, and through your ears, she has whispered to the human mind, the favorite daughter of God the Creator, how she exists in the innermost bosom.

(Shall I have committed a crime if I ask the single composers of this generation for some artistic motet instead of this epigraph? The Royal Psalter and the other Holy Books can supply a text suited for this. But alas for you! No more than six are in concord in the heavens. For the moon sings here monody separately, like a dog sitting on the Earth. Compose the melody; I, in order that the book may progress, promise that I will watch carefully over the six parts. To him who more properly expresses the celestial music described in this work, Clio will give a garland, and Urania will betroth Venus his bride.)

It has been unfolded above what harmonic ratios two neighbouring planets would embrace in their extreme movements. But it happens very rarely that two, especially the slowest, arrive at their extreme intervals at the same time; for example, the apsides of Saturn and Jupiter are about 81° apart. Accordingly, while this distance between them measures out the whole zodiac by definite twenty-year leaps,[2] eight hundred years pass by, and nonetheless the leap which concludes the eighth century, does not

---

1. See note on hexachordal system.
2. That is to say, since Saturn and Jupiter have one revolution with respect to one another every twenty years, they are 81° apart once every twenty years, while the end-positions of this 81° interval traverse the ecliptic in leaps, so to speak, and coincide with the apsides approximately once in eight hundred years. C. G. W.

carry precisely to the very apsides; and if it digresses much further, another eight hundred years must be awaited, that a more fortunate leap than that one may be sought; and the whole route must be repeated as many times as the measure of digression is contained in the length of one leap. Moreover, the other single pairs of planets have periods as that, although not so long. But meanwhile there occur also other consonances of two planets, between movements whereof not both are extremes but one or both are intermediate; and those consonances exist as it were in different tunings (*tensionibus*). For, because Saturn tends from *G* to *b*, and slightly further, and Jupiter from *b* to *d* and further; therefore between Jupiter and Saturn there can exist the following consonances, over and above the octave: the major and minor third and the perfect fourth, either one of the thirds through the tuning which maintains the amplitude of the remaining one, but the perfect fourth through the amplitude of a major whole tone. For there will be a perfect fourth not merely from *G* of Saturn to *cc* of Jupiter but also from *A* of Saturn to *dd* of Jupiter and through all the intermediates between the *G* and *A* of Saturn and the *cc* and *dd* of Jupiter. But the octave and the perfect fifth exist solely at the points of the apsides. But Mars, which got a greater interval as its own, received it in order that it should also make an octave with the upper planets through some amplitude of tuning. Mercury received an interval great enough for it to set up almost all the consonances with all the planets within one of its periods, which is not longer than the space of three months. On the other hand, the Earth, and Venus much more so, on account of the smallness of their intervals, limit the consonances, which they form not merely with the others but with one another in especial, to visible fewness. But if three planets are to concord in one harmony, many periodic returns are to be awaited; nevertheless there are many consonances, so that they may so much the more easily take place, while each nearest consonance follows after its neighbour, and very often threefold consonances are seen to exist between Mars, the Earth, and Mercury. But the consonances of four planets now begin to be scattered throughout centuries, and those of five planets throughout thousands of years.

But that all six should be in concord [296] has been fenced about by the longest intervals of time; and I do not know whether it is absolutely impossible for this to occur twice by precise evolving or whether that points to a certain beginning of time, from which every age of the world has flowed.

But if only one sextuple harmony can occur, or only one notable one among many, indubitably that could be taken as a sign of the Creation. Therefore we must ask, in exactly how many forms are the movements of all six planets reduced to one common harmony? The method of inquiry is as follows: let us begin with the Earth and Venus, because these two planets do not make more than two consonances and (wherein the cause of this thing is comprehended) by means of very short intensifications of the movements.

Therefore let us set up two, as it were, skeletal outlines of harmonies, each skeletal outline determined by the two extreme numbers wherewith the limits of the tunings are designated, and let us search out what fits in with them from the variety of movements granted to each planet.

HARMONIES OF ALL THE PLANETS, OR UNIVERSAL HARMONIES IN THE MAJOR MODE

In order that b may be in concord			At gravest Tuning	At most acute Tuning	[Modern notation
			380'20"		
☿	e⁷		285'15"	292'48"	5 x 8va
	b⁶		228'12"	234'16"	
	g⁶				
♀	e⁶		190'10"	195'14"	4 x 8va
	e⁵		95'5"	97'37"	
☊	g⁴		57'3"	58'34"	2 x 8va
	b³		35'39"	36'36"	
♂	g³				8va
			28'32"	29'17"	
♃	b			4'34"	
♭	B		2'14"		
	G		1'47"	1'49"	

E. C., Jr.]

In order that c may be in concord			At gravest Tuning	At most acute Tuning	[Modern notation
☿	e⁷		380'20"		5 x 8va
	c⁷		204'16"	312'21"	
	g⁶		228'12"	234'16"	
♀	e⁶		190'10"	195'14"	4 x 8va
	e⁵		95'5"	97'37"	
☊	g⁴		57'3"	58'34"	2 x 8va
	c⁴		38'2"	39'3"	

In order that c *may be in concord*	At gravest Tuning	At most acute Tuning	[Modern notation
♂⁷ g⁸	28'32"	29'17"	8va
♃ c¹	4'45"	4'53"	
♭ G	1'47"	1'49"	E. C., Jr.]

E. C., Jr.]

Saturn joins in this universal consonance with its aphelial movement, the Earth with its aphelial, Venus approximately with its aphelial; at highest tuning, Venus joins with its perihelial; at mean tuning, Saturn joins with its perihelial, Jupiter with its aphelial, Mercury with its perihelial. So Saturn can join in with two movements, Mars with two, Mercury with four. But with the rest remaining, the perihelial movement of Saturn and the aphelial of Jupiter are not allowed. But in their place, Mars joins in with perihelial movement.

The remaining planets join in with single movements, Mars alone with two, and Mercury with four.

[297] Accordingly, the second skeletal outline will be that wherein the other possible consonance, 5 : 8, exists between the Earth and Venus. Here one eighth of the 94'50" of the diurnal aphelial movement of Venus or 11'51"+, if multiplied by 5, equals the 59'16" of the movement of the Earth; and similar parts of the 97'37" of the perihelial movement of Venus are equal to the 61'1" of the movement of the Earth. Accordingly, the other planets are in concord in the following diurnal movements:

HARMONIES OF ALL THE PLANETS, OR UNIVERSAL HARMONIES
IN THE MINOR MODE

In order that b *may be in concord*	At gravest Tuning	At most acute Tuning	[Modern notation
♄ e♭⁷ b♭⁷ g⁶	379'20" 284'32" 237'4"	295'56" 244'4"	5 x 8va
♀ e♭⁶ e♭⁵	189'40" 94'50"	195'14" 97'37"	4 x 8va

In order that c may be in concord	At gravest Tuning	At most acute Tuning	[Modern notation
♄ g⁴ / b♭⁴	59'16" / 35'35"	61'1" / 36'37"	2 x 8va
♂ g³	29'38"	30'31"	8va
♃ b♭¹		4'35"	
♄ b / G	2'13" / 3'51"	1'55"	

E. C., Jr.]

Here again, in the mean tuning Saturn joins in with its perihelial movement, Jupiter with its aphelial, Mercury with its perihelial. But at highest tuning approximately the perihelial movement of the Earth joins in.

In order that c may be in concord	At gravest Tuning	At most acute Tuning	[Modern notation
☿ e♭⁷ / c⁷ / g⁶	379'20" / 316'5" / 237'4"	325'26" / 244'4"	5 x 8va
♀ e♭⁶ / c⁶ / e♭⁵	189'40" / 94'50"	195'14" / 162'43" / 97'37"	4 x 8va
♄ g⁴	59'16"	61'1"	2 x 8va
♂ g³	29'38"	30'31"	8va
♃ c¹	4'56"	5'5"	
♄ G	3'51"	1'55"	

E. C., Jr.]

And here, with the aphelial movement of Jupiter and the perihelial movement of Saturn removed, the aphelial movement of Mercury is practically admitted besides the perihelial. The rest remain.

Therefore astronomical experience bears witness that the universal consonances of all the movements can take place, and in the two modes (*generum*), the major and minor, and in both genera of form, or (if I may say so) in respect to two pitches and in any one of the four cases, with a certain latitude of tuning and also with a certain variety in the particular consonances of Saturn, Mars, and Mercury, of each with the rest; and that is not afforded by the intermediate movements alone, but by all the extreme movements too, except the aphelial movement of Mars and the perihelial movement of Jupiter; because since the former occupies *f* sharp; and the latter, *d* Venus, which occupies perpetually the intermediate *e* flat or *e*, does not allow those neighbouring dissonances in the universal consonance, as she would do if she had space to go beyond *e* or *e* flat. This difficulty is caused by the wedding of the Earth and Venus, or the male and the female. These two planets divide the kinds (*genera*) of consonances into the major and masculine and the minor and feminine, according as the one spouse has gratified the other—namely, either the Earth is in its aphelion, as if preserving [298] its marital dignity and performing works worthy of a man, with Venus removed and pushed away to her perihelion as to her distaff; or else the Earth has kindly allowed her to ascend into aphelion or the Earth itself has descended into its perihelion towards Venus and as it were, into her embrace, for the sake of pleasure, and has laid aside for a while its shield and arms and all the works befitting a man; for at that time the consonance is minor.

But if we command this contradictory Venus to keep quiet, *i.e.*, if we consider what the consonances not of all but merely of the five remaining planets can be, excluding the movement of Venus, the Earth still wanders around its g string and does not ascend a semitone above it. Accordingly *b*-flat, *b*, *c*, *d*, *e*-flat, and *e* can be in concord with *g*, whereupon, as you see, Jupiter, marking the d string with its perihelial movement, is brought in. Accordingly, the difficulty about Mars' aphelial movement remains. For the aphelial movement of the Earth, which occupies *g*, does not allow it on *f* sharp; but the perihelial movement, as was said above in Chapter V, is in discord with the aphelial movement of Mars by about half a diesis.

HARMONIES OF THE FIVE PLANETS, WITH VENUS LEFT OUT

*Major mode (*Genus durum*)*		At gravest Tuning	At most acute Tuning	[Modern notation
d⁷ b⁶ g⁶		342'18" 285'15" 228'12"	351'24" 292'48" 234'16"	5 x 8va
♀ in d⁶ discord e⁵		171'9" 95'5"	175'42" 97'37"	4 x 8va

42

Major mode (Genus durum)		At gravest Tuning	At most acute Tuning	[Modern notation
♂	g⁴ b³	57'3" 35'39"	58'34" 36'36"	2 x 8va
	g³	28'31"	29'17"	8va
♃	d¹ b¹	5'21"	5'30" 4'35"	
♭	B G	2'13" 1'47"		

E. C., Jr.]

Here at the most grave tuning, Saturn and the Earth join in with their aphelial movements; at the mean tuning, Saturn with its perihelial and Jupiter with its aphelial; at the most acute, Jupiter with its perihelial.

Minor mode (Genus molle)		At gravest tuning	At most acute tuning	[Modern notation:
☿	d⁷ b⁶ g⁶	342'18" 273'50" 228'12"	351'24" 280'57" 234'16"	5 x 8va
♀ in discord	d⁶ e⁵	171'9" 95'5"	175'42" 97'37"	4 x 8va
♂	g⁴ b³	57'3" 34'14"	58'34" 35'8"	2 x 8va
	g³	28'31"	29'17"	8va
♃	d¹	5'21"	5'30"	
♭	B G	2'8" 1'47"	2'12" 1'50"	

E. C., Jr.]

Here the aphelial movement of Jupiter is not allowed, but at the most acute tuning Saturn practically joins in with its perihelial movement.

But there can also exist the following harmony of the four planets, Saturn, Jupiter, Mars, and Mercury, wherein too the aphelial movement of Mars is present, but it is without latitude of tuning.

In order that b *may be in concord* | *[Modern notation:*

d⁷ / b⁶	335'50" / 279'52"
f#⁶ / d⁶	209'52" / 167'55"
♂ b³	34'59"
f#³	26'14"
♃ d¹	5'15"
♭ B	2'11"

E. C., Jr.]

In order that a *may be in concord* | *[Modern notation:*

44

In order that a *may be in concord* [Modern notation:

E. C., Jr.]

Accordingly the movements of the heavens are nothing except a certain everlasting polyphony (intelligible, not audible) with dissonant tunings, like certain syncopations or cadences (wherewith men imitate these natural dissonances), which tends towards fixed and prescribed clauses—the single clauses having six terms (like voices)—and

1. The comparison Kepler draws between the celestial harmonies and the polyphonic music of his time may be clarified by a simple example for four voices from—Palestrina, *O Crux:*

X   Consonant harmonies
Y   Dissonant syncopations
Z   Resolutions of dissonances

As will be observed each of the few voices (as it would also be with the six to which Kepler refers) moves from one consonant chord to another while following a graceful melodic line. Sometimes bits of scales or passing tones are added to give a voice more melodic freedom expressiveness. For the same reason a voice may remain on the same note while the other voices change to a new chord. When this becomes a dissonance (called a syncopation) in the new chord it usually resolves by moving one step downward to a tone that is consonant with the other voices. As in this example each section or "cause" ends with a cadence.

E. C., Jr.

which marks out and distinguishes the immensity of time with those notes.[1] Hence it is no longer a surprise that man, the ape of his Creator, should finally have discovered the art of singing polyphonically (*per concentum*), which was unknown to the ancients, namely in order that he might play the everlastingness of all created time in some short part of an hour by means of an artistic concord of many voices and that he might to some extent taste the satisfaction of God the Workman with His own works, in that very sweet sense of delight elicited from this music which imitates God.

8. IN THE CELESTIAL HARMONIES WHICH PLANET SINGS SOPRANO, WHICH ALTO, WHICH TENOR, AND WHICH BASS?

Although these words are applied to human voices, while voices or sounds do not exist in the heavens, on account of the very great tranquillity of movements, and not even the subjects in which we find the consonances are comprehended under the true genus of movement, since we were considering the movements solely as apparent from the sun, and finally, although there is no such cause in the heavens, as in human singing, for requiring a definite number of voices in order to make consonance (for first there was the number of the six planets revolving around the sun, from the number of the five intervals taken from the regular figures, and then afterwards—in the order of nature, not of time—the congruence of the movements was settled): I do not know why but nevertheless this wonderful congruence with human song has such a strong effect upon me that I am compelled to pursue this part of the comparison, also, even without any solid natural cause. For those same properties which in Book III, [300] Chapter 16, custom ascribed to the bass and nature gave legal grounds for so doing are somehow possessed by Saturn and Jupiter in the heavens; and we find those of the tenor in Mars, those of the alto are present in the Earth and Venus, and those of the soprano are possessed by Mercury, if not with equality of intervals, at least proportionately. For howsoever in the following chapter the eccentricities of each planet are deduced from their proper causes and through those eccentricities the intervals proper to the movements of each, none the less there comes from that the following wonderful result (I do not know whether it is occasioned by the procurement and mere tempering of necessities): (1) as the bass is opposed to the alto, so there are two planets which have the nature of the alto, two that of the bass, just as in any Mode of song there is one (bass and one alto) on either side, while there are single representatives of the other single voices. (2) As the alto is practically supreme in a very narrow range (*in angustiis*) on account of necessary and natural causes unfolded in Book III, so the almost innermost planets, the Earth and Venus, have the narrowest intervals of movements, the Earth not much more than a semitone, Venus not even a diesis. (3) And as the tenor is free, but none the less progresses with moderation, so Mars alone—with the single exception of Mercury—can make the greatest interval, namely a perfect fifth. (4) And as the bass makes harmonic leaps, so Saturn and Jupiter have intervals which are harmonic, and in relation to one another pass from the octave to the octave and perfect fifth. (5) And as the soprano is the freest, more than all the rest, and likewise the swiftest, so Mercury can traverse more than an octave in the shortest period. But this is altogether *per accidens*; now let us hear the reasons for the eccentricities.

9. THE GENESIS OF THE ECCENTRICITIES IN THE SINGLE PLANETS FROM THE PROCUREMENT OF THE CONSONANCES BETWEEN THEIR MOVEMENTS

Accordingly, since we see that the universal harmonies of all six planets cannot take place by chance, especially in the case of the extreme movements, all of which we see concur in the universal harmonies—except two, which concur in harmonies closest to the universal—and since much less can it happen by chance that all the pitches of the system of the octave (as set up in Book III) by means of harmonic divisions are designated by the extreme planetary movements, but least of all that the very subtle business of the distinction of the celestial consonances into two modes, the major and minor, should be the outcome of chance, without the special attention of the Artisan: accordingly it follows that the Creator, the source of all wisdom, the everlasting approver of order, the eternal and superexistent geyser of geometry and harmony, it follows, I say, that He, the Artisan of the celestial movements Himself, should have conjoined to the five regular solids the harmonic ratios arising from the regular plane figures, and out of both classes should have formed one most perfect archetype of the heavens: in order that in this archetype, as through the five regular solids the shapes of the spheres shine through on which the six planets are carried, so too through the consonances, which are generated from the plane figures, and deduced from them in Book III, the measures of the eccentricities in the single planets might be determined so as to proportion the movements of the planetary bodies; and in order that there should be one tempering together of the ratios and the consonances, and that the greater ratios of the spheres should yield somewhat to the lesser ratios of the eccentricities necessary for procuring the consonances, and conversely those in especial of the harmonic ratios which had a greater kinship with each solid figure should be adjusted to the planets—in so far as that could be effected by means of consonances. And in order that, finally, in that way both the ratios of the spheres and the eccentricities of the single planets might be born of the archetype simultaneously, while from the amplitude of the spheres and the bulk of the bodies the periodic times of the single planets might result.

[301] While I struggle to bring forth this process into the light of human intellect

47

by means of the elementary form customary with geometers, may the Author of the heavens be favourable, the Father of intellects, the Bestower of mortal senses, Himself immortal and superblessed, and may He prevent the darkness of our mind from bringing forth in this work anything unworthy of His Majesty, and may He effect that we, the imitators of God by the help of the Holy Ghost, should rival the perfection of His works in sanctity of life, for which He choose His church throughout the Earth and, by the blood of His Son, cleansed it from sins, and that we should keep at a distance all the discords of enmity, all contentions, rivalries, anger, quarrels, dissensions, sects, envy, provocations, and irritations arising through mocking speech and the other works of the flesh; and that along with myself, all who possess the spirit of Christ will not only desire but will also strive by deeds to express and make sure their calling, by spurning all crooked morals of all kinds which have been veiled and painted over with the cloak of zeal or of the love of truth or of singular erudition or modesty over against contentious teachers, or with any other showy garment. Holy Father, keep us safe in the concord of our love for one another, that we may be one, just as Thou art one with Thine Son, Our Lord, and with the Holy Ghost, and just as through the sweetest bonds of harmonies Thou hast made all Thy works one; and that from the bringing of Thy people into concord the body of Thy Church may be rebuilt up in the Earth, as Thou didst erect the heavens themselves out of harmonies.

PRIOR REASONS

I. Axiom. *It is reasonable that, wherever in general it could have been done, all possible harmonies were due to have been set up between the extreme movements of the planets taken singly and by twos, in order that that variety should adorn the world.*

II. Axiom. *The five intervals between the six spheres to some extent were due to correspond to the ratio of the geometrical spheres which inscribe and circumscribe the five regular solids, and in the same order which is natural to the figures.*

Concerning this, see Chapter 1 and the *Mysterium Cosmographicum* and the *Epitome of Copernican Astronomy.*

III. Proposition. *The intervals between the Earth and Mars, and between the Earth and Venus, were due to be least, in proportion to their spheres, and thereby approximately equal; middling and approximately equal between Saturn and Jupiter, and between Venus and Mercury; but greatest between Jupiter and Mars.*

For by Axiom II, the planets corresponding in position to the figures which make the least ratio of geometrical spheres ought likewise to make the least ratio; but those which correspond to the figures of middling ratio ought to make the greatest; and those which correspond to the figures of greatest ratio, the greatest. But the order holding between the figures of the dodecahedron and the icosahedron is the same as that between the pairs of planets, Mars and the Earth, and the Earth and Venus, and the order of the cube and octahedron is the same as that of the pair Saturn and Jupiter and that of the pair Venus and Mercury; and, finally, the order of the tetrahedron is the same as that of the pair Jupiter and Mars (see Chapter 3). Therefore, the least ratio will hold between the planetary spheres first mentioned, while that between Saturn and Jupiter is approximately equal to that between Venus and Mercury; and, finally, the greatest between the spheres of Jupiter and Mars.

IV. Axiom. *All the planets ought to have their eccentricities diverse, no less than a movement in latitude, and in proportion to those eccentricities also their distances from the sun, the source of movement, diverse.*

As the essence of movement consists not in *being* but in *becoming*, so too the form or figure of the region which any planet traverses in its movement does not become solid immediately from the start but in the succession of time acquires at last not only length but also breadth and depth (its perfect ternary of dimensions); and, gradually, thus, by the interweaving and piling up of many circuits, the form of a concave sphere comes to be represented—just as out of the silk-worm's thread, by the interweaving and heaping together of many circles, the cocoon is built.

V. Proposition. *Two diverse consonances were to have been attributed to each pair of neighbouring planets.*

For, by Axiom IV, any planet has a longest and a shortest distance from the sun, wherefore, by Chapter 3, it will have both a slowest movement and a fastest. Therefore, there are two primary comparisons of the extreme movements, one of the diverging movements in the two planets, and the other of the converging. Now it is necessary that they be diverse from one another, because the ratio of the diverging movements will be greater, that of the converging, lesser. But, moreover, diverse consonances had to exist by way of diverse pairs of planets, so that this variety should make for the adornment of the world—by Axiom I—and also because the ratios of the intervals between two planets are diverse, by Proposition III. But to each definite ratio of the spheres there correspond harmonic ratios, in quantitative kinship, as has been demonstrated in Chapter 5 of this book.

VI. Proposition. *The two least consonances, 4 : 5 and 5 : 6, do not have a place between two planets.*

For

$$5 : 4 = 1,000 : 800$$

and

$$6 : 5 = 1,000 : 833.$$

But the spheres circumscribed around the dodecahedron and icosahedron have a greater ratio to the inscribed spheres than 1,000 : 795, etc., and these two ratios indicate the intervals between the nearest planetary spheres, or the least distances. For in the other regular solids the spheres are farther distant from one another. But now the ratio of the movements is even greater than the ratios of the intervals, unless the ratio of the eccentricities to the spheres is vast—by Article XIII of Chapter 3. Therefore the least ratio of the movements is greater than 4 : 5 and 5 : 6. Accordingly, these consonances, being hindered by the regular solids, receive no place among the planets.

VII. Proposition. *The consonance of the perfect fourth can have no place between the converging movements of two planets, unless the ratios of the extreme movements proper to them are, if compounded, more than a perfect fifth.*

For let 3 : 4 be the ratio between the converging movements. And first, let there be no eccentricity, no ratio of movements proper to the single planets, but both the converging and the mean movements the same; then it follows that the corresponding intervals, which by this hypothesis will be the semidiameters of the spheres, constitute the $^2/_3$d power of this ratio, *viz.*, 4480 : 5424 (by Chapter 3). But this ratio is already less than the ratio of the spheres of any regular figure; and so the whole inner sphere would be cut by the regular planes of the figure inscribed in any outer sphere. But this is contrary to Axiom II.

Secondly, let there be some composition of the ratios between the extreme movements, and let the ratio of the converging movements be 3 : 4 or 75 : 100, but let the ratio of the corresponding intervals be 1,000 : 795, since no regular figure has a lesser ratio of spheres. And because the inverse ratio of the movements exceeds this ratio of the intervals by the excess 750 : 795, then if this excess is divided into the ratio 1,000 : 795, according to the doctrine of Chapter 3, the result will be 9434 : 7950, the square root of the ratio of the spheres. Therefore the square of this ratio, *viz.*, 8901 : 6320, *i.e.*, 10,000 : 7,100 is the ratio of the spheres. Divide this by 1000 : 795, the ratio of the converging intervals, the result will be 7100 : 7950, about a major whole tone. The compound of the two ratios which the mean movements have to the converging movements on either side must be at least so great, in order that the perfect fourth may be possible between the converging movements. Accordingly, the compound ratio of the diverging extreme intervals to the converging extreme intervals is about the square root of this ratio, *i.e.*, two tones, and again the converging intervals are the square of this, *i.e.*, more than a perfect fifth. Accordingly, if the compound of the proper movements of two neighbouring planets is less than a perfect fifth, a perfect fourth will not be possible between their converging movements.

VIII. Proposition. *The consonances 1 : 2 and 1 : 3, i.e., the octave and the octave plus a fifth were due to Saturn and Jupiter.*

For they are the first and highest of the planets and have obtained the first figure, the cube, by Chapter 1 of this book; and these consonances are first in the order of nature and are chief in the two families of figures, the bisectorial or tetragonal and the triangular, by what has been said in Book I. But that which is chief, the octave 1 : 2, is approximately greater than the ratio of the spheres of the cube, [303] which is $1 : \sqrt{3}$; wherefore it is fitted to become the lesser ratio of the movements of the planets on the cube, by Chapter 3, Article XIII; and, as a consequence, 1 : 3 serves as the greater ratio.

But this is also the same as what follows: for if some consonance is to some ratio of the spheres of the figures, as the ratio of the movements apparent from the sun is to the ratio of the mean intervals, such a consonance will duly be attributed to the movements. But it is natural that the ratio of the diverging movements should be much greater than the ratio of the $^3/_2$th powers of the spheres, according to the end of Chapter 3, *i.e.*, it approaches the square of the ratio of the spheres; and moreover 1 : 3 is the square of the ratio of the spheres of the cube, which we call the ratio of $1 : \sqrt{3}$. Therefore, the ratio of the diverging movements of Saturn and Jupiter is 1 : 3. (See above, Chapter 2, for many other kinships of these ratios with the cube.)

IX. Proposition. *The private ratios of the extreme movements of Saturn and Jupiter compounded were due to be approximately 2 : 3, a perfect fifth.*

This follows from the preceding; if the perihelial movement of Jupiter is triple the aphelial movement of Saturn, and conversely the aphelial movement of Jupiter is double the perihelial of Saturn, then 1 : 2 and 1 : 3 compounded inversely give 2 : 3.

X. Axiom. *When choice is free in other respects, the private ratio of movements, which is prior in nature or of a more excellent mode or even which is greater, is due to the higher planet.*

XI. Proposition. *The ratio of the aphelial movement of Saturn to the perihelial was due to be 4 : 5, a major third, but that of Jupiter's movements 5 : 6, a minor third.*

For as compounded together they are equivalent to 2 : 3; but 2 : 3 can be divided harmonically no other way than into 4 : 5 and 5 : 6. Accordingly God the composer of harmonies divided harmonically the consonance 2 : 3, (by Axiom I) and the harmonic part of it which is greater and of the more excellent major mode, as masculine, He gave to Saturn the greater and higher planet, and the lesser ratio 5 : 6 to the lower one, Jupiter (by Axiom X).

XII. Proposition. *The great consonance of 1 : 4, the double octave, was due to Venus and Mercury.*

For as the cube is the first of the primary figures, so the octahedron is the first of the secondary figures, by Chapter 1 of this book. And as the cube considered geometrically is outer and the octahedron is inner, *i.e.*, the latter can be inscribed in the former, so also in the world Saturn and Jupiter are the beginning of the upper and outer planets, or from the outside; and Mercury and Venus are the beginning of the inner planets, or from the inside, and the octahedron has been placed between their circuits:

(see Chapter 3). Therefore, from among the consonances, one which is primary and cognate to the octahedron is due to Venus and Mercury. Furthermore, from among the consonances, after 1 : 2 and 1 : 3, there follows in natural order 1 : 4; and that is cognate to 1 : 2, the consonance of the cube, because it has arisen from the same cut of figures, *viz.*, the tetragonal, and is commensurable with it, *viz.*, the double of it; while the octahedron is also akin to, and commensurable with the cube. Moreover, 1 : 4 is cognate to the octahedron for a special reason, on account of the number four being in that ratio, while a quadrangular figure lies concealed in the octahedron and the ratio of its spheres is said to be $1 : \sqrt{2}$.

Accordingly the consonance 1 : 4 is a continued power of this ratio, in the ratio of the squares, *i.e.*, the 4th power of $1 : \sqrt{2}$ (see Chapter 2). Therefore, 1 : 4 was due to Venus and Mercury. And because in the cube 1 : 2 has been made the smaller consonance of the two, since the outermost position is over against it, in the octahedron there will be 1 : 4, the greater consonance of the two, as the innermost position is over against it. But too, this is the reason why 1 : 4 has here been given as the greater consonance, not as the smaller.[1] For since the ratio of the spheres of the octahedron is the ratio of $1 : \sqrt{3}$, then if it is postulated that the inscription of the octahedron among the planets is perfect (although it is not perfect, but penetrates Mercury's sphere to some extent—which is of advantage to us): accordingly, the ratio of the converging movements must be less than the $^3/_2$th powers of $1 : \sqrt{3}$; but indeed 1 : 3 is plainly the square of the ratio $1 : \sqrt{3}$ and is thus greater than the exact ratio; all the more then will 1 : 4 be greater than the exact ratio, as greater than 1 : 3. Therefore, not even the square root of 1 : 4 is allowed between the converging movements. Accordingly, 1 : 4 cannot be less than the octahedric; so it will be greater.

Further: 1 : 4 is akin to the octahedric square, where the ratio of the inscribed and circumscribed circles is $1 : \sqrt{2}$, just as 1 : 3 is akin to the cube, where the ratio of the spheres is $1 : \sqrt{3}$. For as 1 : 3 is a power of $1 : \sqrt{3}$, *viz.*, its square, [304] so too here 1 : 4 is a power of $1 : \sqrt{2}$, *viz.*, twice its square, *i.e.*, its quadruple power. Wherefore, if 1 : 3 was due to have been the greater consonance of the cube (by Proposition VII), accordingly 1 : 4 ought to become the greater consonance of its octahedron.

XIII. Proposition. *The greater consonance of approximately 1 : 8, the triple octave, and the smaller consonance of 5 : 24, the minor third and double octave, were due to the extreme movements of Jupiter and Mars.*

For the cube has obtained 1 : 2 and 1 : 3, while the ratio of the spheres of the tetrahedron, which is situated between Jupiter and Mars, called the triple ratio, is the square of the ratio of the spheres of the cube, which is called the ratio of $1 : \sqrt{3}$. Therefore, it was proper that ratios of movements which are the squares of the cubic

---

1. *Smaller* (lesser) and *greater* consonances are equivalent to our modern "more closely spaced" and "more widely spaced" consonances. E. C., Jr.

ratios should be applied to the tetrahedron. But of the ratios 1 : 2 and 1 : 3 the following ratios are the squares: 1 : 4 and 1 : 9. But 1 : 9 is not harmonic, and 1 : 4 has already been used up in the octahedron. Accordingly, consonances neighbouring upon these ratios were to have been taken, by Axiom I. But the lesser ratio 1 : 8 and the greater 1 : 10 are the nearest. Choice between these ratios is determined by kinship with the tetrahedron, which has nothing in common with the pentagon, since 1 : 10 is of a pentagonal cut, but the tetrahedron has greater kinship with 1 : 8 for many reasons (see Chapter 2).

Further, the following also makes for 1 : 8: just as 1 : 3 is the greater consonance of the cube and 1 : 4 the greater consonance of the octahedron, because they are powers of the ratios between the spheres of the figures, so too 1 : 8 was due to be the greater consonance of the tetrahedron, because as its body is double that of the octahedron inscribed in it, as has been said in Chapter 1, so too the term 8 in the tetrahedral ratio is double the term 4 in the tetrahedral ratio.

Further, just as 1 : 2 the smaller consonance of the cube, is one octave, and 1 : 4, the greater consonance of the octahedron, is two octaves, so already 1 : 8, the greater consonance of the tetrahedron, was due to be three octaves. Moreover, more octaves were due to the tetrahedron than to the cube and octahedron, because, since the smaller tetrahedral consonance is necessarily greater than all the lesser consonances in the other figures (for the ratio of the tetrahedral spheres is greater than all the spheres of figures): too the greater tetrahedral consonance was due to exceed the greater consonances of the others in number of octaves. Finally, the triple of octave intervals has kinship with the triangular form of the tetrahedron, and has a certain perfection, as follows: every three is perfect; since even the octuple, the term (of the triple octave), is the first cubic number of perfect quantity, namely of three dimensions.

A greater consonance neighbouring upon 1 : 4 or 6 : 24 is 5 : 24, while a lesser is 6 : 20 or 3 : 10. But again 3 : 10 is of the pentagonal cut, which has nothing in common with the tetrahedron. But on account of the numbers 3 and 4 (from which the numbers 12, 24 arise) 5 : 24 has kinship with the tetrahedron. For we are here neglecting the other lesser terms, *viz.*, 5 and 3, because their lightest degree of kinship is with figures, as it is possible to see in Chapter 2. Moreover, the ratio of the spheres of the tetrahedron is triple; but the ratio of the converging intervals too ought to be approximately so great, by Axiom II. By Chapter 3, the ratio of the converging movements approaches the inverse ratio of the $^3/_2$th powers of the intervals, but the $^3/_2$th power of 3 : 1 is approximately 1000 : 193. Accordingly, whereof the aphelial movement of Mars is 1000, the (perihelial) of Jupiter will be slightly greater than 193 but much less than 333, which is one third of 1,000. Accordingly, not the consonance 10 : 3, *i.e.*, 1,000 : 333, but the consonance 24 : 5, *i.e.*, 1,000 : 208, takes place between the converging movements of Jupiter and Mars.

XIV. Proposition. *The private ratio of the extreme movements of Mars was due to be greater than 3 : 4, the perfect fourth, and approximately 18 : 25.*

For let there be the exact consonances 5 : 24 and 1 : 8 or 3 : 24, which are commonly attributed to Jupiter and Mars (Proposition XIII). Compound inversely 5 : 24, the lesser with 3 : 24, the greater; 3 : 5 results as the compound of both ratios. But the proper ratio of Jupiter alone has been found to be 5 : 6, in Proposition XI, above. Then compound this inversely with the composition 3 : 5, *i.e.*, compound 30 : 25 and 18 : 30; there results as the proper ratio of Mars 18 : 25, which is greater than 18 : 24 or 3 : 4. But it will become still greater, if, on account of the ensuing reasons, the common greater consonance 1 : 8 is increased.

XV. Proposition. *The consonances 2 : 3, the fifth; 5 : 8, the minor sixth; and 3 : 5, the major sixth were to have been distributed among the converging movements of Mars and the Earth, the Earth and Venus, Venus and Mercury, and in that order.*

For the dodecahedron and the icosahedron, the figures interspaced between Mars, the Earth, and Venus have the least ratio between their circumscribed and inscribed spheres. [305] Therefore from among possible consonances the least are due to them, as being cognate for this reason, and in order that Axiom II may have place. But the least consonances of all, *viz.*, 5 : 6 and 4 : 5, are not possible, by Proposition IV. Therefore, the nearest consonances greater than they, *viz.*, 3 : 4 or 2 : 3 or 5 : 8 or 3 : 5 are due to the said figures.

Again, the figure placed between Venus and Mercury, *viz.*, the octahedron, has the same ratio of its spheres as the cube. But by Proposition VII, the cube received the octave as the lesser consonance existing between the converging movements. Therefore, by proportionality, so great a consonance, *viz.*, 1 : 2, would be due to the octahedron as the lesser consonance, if no diversity intervened. But the following diversity intervenes: if compounded together, the private ratios of the single movements of the cubic planets, *viz.*, Saturn and Jupiter, did not amount to more than 2 : 3; while, if compounded, the ratios of the single movements of the octahedral planets, *viz.*, Venus and Mercury will amount to more than 2 : 3, as is apparent easily, as follows: For, as the proportion between the cube and octahedron would require if it were alone, let the lesser octahedral ratio be greater than the ratios here given, and thereby clearly as great as was the cubic ratio, *viz.*, 1 : 2; but the greater consonance was 1 : 4, by Proposition XII. Therefore if the lesser consonance 1 : 2 is divided into the one we have just laid down, 1 : 2, still remains as the compound of the proper movements of Venus and Mercury; but 1 : 2 is greater than 2 : 3 the compound of the proper movements of Saturn and Jupiter; and indeed a greater eccentricity follows upon this greater compound, by Chapter 3, but a lesser ratio of the converging movements follows upon the greater eccentricity, by the same Chapter 3. Wherefore by the addition of a greater eccentricity to the proportion between the cube and the octahedron it comes about that a lesser ratio than 1 : 2 is also required between the converging movements of Venus and Mercury. Moreover, it was in keeping with Axiom I that, with the consonance of the octave given to the planets of the cube, another consonance which is very near (and by the earlier demonstration less than 1 : 2) should be joined to the planets of the octahedron. But 3 : 5 is proximately less than 1 : 2, and as the greatest of the three it was due to the figure having the greatest ratio of its spheres, *viz.*, the octahedron. Accordingly, the lesser ratios, 5 : 8 and 2 : 3 or 3 : 4, were left for the icosahedron and dodecahedron, the figures having a lesser ratio of their spheres.

But these remaining ratios have been distributed between the two remaining planets, as follows. For as, from among the figures, though of equal ratios between their spheres, the cube has received the consonance 1 : 2, while the octahedron the lesser consonance 3 : 5, in that the compound ratio of the private movements of Venus and Mercury exceeded the compound ratio of the private movements of Saturn and Jupiter; so also although the dodecahedron has the same ratio of its spheres as the icosahedron, a lesser ratio was due to it than to the icosahedron, but very close on account of a similar reason, *viz.*, because this figure is between the Earth and Mars, which had a great eccentricity in the foregoing. But Venus and Mercury, as we shall hear in the following, have the least eccentricities. But since the octahedron has 3 : 5, the icosahedron, whose species are in a lesser ratio, has the next slightly lesser, *viz.*, 5 : 8; accordingly, either 2 : 3, which remains, or 3 : 4 was left for the dodecahedron, but more likely 2 : 3, as being nearer to the icosahedral 5 : 8; since they are similar figures.

But 3 : 4 indeed was not possible. For although, in the foregoing, the private ratio of the extreme movements of Mars was great enough, yet the Earth—as has already been said and will be made clear in what follows—contributed its own ratio, which was too small for the compound ratio of both to exceed the perfect fifth. Accordingly, Proposition VII, 3 : 4 could not have place. And all the more so, because—as will follow in Proposition XVII—the ratio of the converging intervals was due to be greater than 1,000 : 795.

XVI. Proposition. *The private ratios of movements of Venus and Mercury, if compounded together, were due to make approximately 5 : 12.*

For divide the lesser harmonic ratio attributed in Proposition XV to this pair jointly into the greater of them, 1 : 4 or 3 : 12, by Proposition XII; there results 5 : 12, the compound ratio of the private movements of both. And so the private ratio of the extreme movements of Mercury alone is less than 5 : 12, the magnitude of the private movement of Venus. Understand this of these first reasons. For below, by the second reasons, through the addition of some variation to the joint consonances of both, it results that only the private ratio of Mercury is perfectly 5 : 12.

XVII. Proposition. *The consonance between the diverging movements of Venus and the Earth could not be less than 5 : 12.*

For in the private ratio of its movements Mars alone has received more than the perfect fourth and more than 18 : 25, by Proposition XIV. But their lesser consonance is the perfect fifth, [306] by Proposition XV. Accordingly, the ratio compounded of these two parts is 12 : 25. But its own private ratio is due to the Earth, by Axiom IV. Therefore, since the consonance of the diverging movements is made up out of the said three elements, it will be greater than 12 : 25. But the nearest consonance greater than 12 : 25, *i.e.,* 60 : 125, is 5 : 12, *viz.,* 60 : 144. Wherefore, if there is need of a consonance for this greater ratio of the two planets, by Axiom I, it cannot be less than 60 : 144 or 5 : 12.

Therefore up to now all the remaining pairs of planets have received their two consonances by necessary reasons; the pair of the Earth and Venus alone has as yet been allotted only one consonance, 5 : 8, by the axioms so far employed. Therefore, we must now take a new start and inquire into its remaining consonance, *viz.,* the greater, or the consonance of the diverging movements.

POSTERIOR REASONS

XVIII. Axiom. *The universal consonances of movements were to be constituted by a tempering of the six movements, especially in the case of the extreme movements.*

This is proved by Axiom I.

XIX. Axiom. *The universal consonances had to come out the same within a certain latitude of movements, namely, in order that they should occur the more frequently.*

For if they had been limited to indivisible points of the movements, it could have happened that they would never occur, or very rarely.

XX. Axiom. *As the most natural division of the kinds* (generum) *of consonances is into major and minor, as has been proved in Book 3, so the universal consonances of both kinds had to be procured between the extreme movements of the planets.*

XXI. Axiom. *Diverse species of both kinds of consonances had to be instituted, so that the beauty of the world might well be composed out of all possible forms of variety—and by means of the extreme movements, at least by means of some extreme movements.*

By Axiom I.

XXII. Proposition. *The extreme movements of the planets had to designate pitches or strings* (chordas) *of the octave system, or notes* (claves) *of as musical scale.*

For the genesis and comparison of consonances beginning from one common term has generated the musical scale, or the division of the octave into its pitches or tones (*sonos*), as has been proved in Book 3. Accordingly, since varied consonances between the extremes of movements are required, by Axioms I, XX, and XXI, wherefore the real division of some celestial system or harmonic scale by the extremes of movements is required.

XXIII. Proposition. *It was necessary for there to be one pair of planets, between the movements of which no consonances could exist except the major sixth 3 : 5 and the minor sixth 5 : 8.*

For since the division into kinds of consonances was necessary, by Axiom XX, and by means of the extreme movements at the apsides, by XXII, because solely the extremes, *viz.*, the slowest and the fastest, need the determination of a manager and orderer, the intermediate tensions come of themselves, without any special care, with the passage of the planet from the slowest movement to the fastest: accordingly, this ordering could not take place otherwise than by having the diesis or 24 : 25 designated by the extremes of the two planetary movements, in that the kinds of consonances are distinguished by the diesis, as was unfolded in Book 3.

But the diesis is the difference either between two thirds, 4 : 5 and 5 : 6, or between two sixths, 3 : 5 and 5 : 8, or between those ratios increased by one or more octave intervals. But the two thirds, 4 : 5 and 5 : 6, did not have place between two planets, by Proposition VI, and neither the thirds nor the sixths increased by the interval of an octave have been found, except 5 : 12 in the pair of Mars and the Earth, and still not otherwise than along with the related 2 : 3, and so the intermediate ratios 5 : 8 and 3 : 5 and 1 : 2 were alike admitted. Therefore, it remains that the two sixths, 3 : 5 and 5 : 8, were to be given to one pair of planets. But too the sixths alone were to be granted to the variation of their movements, in such fashion that they would neither expand their term to the proximately greater interval of one octave, 1 : 2, [307] nor contract them to the narrows of the proximately lesser interval of the fifth, 2 : 3. For, although it is true that the same two planets, which make a perfect fifth with their extreme converging movements, can also make sixths and thus traverse the diesis too, still this would not smell of the singular providence of the Orderer of movements. For the diesis, the least interval—which is potentially latent in all the major intervals comprehended by the extreme movements—is itself at that time traversed by the intermediate movements varied by continuous tension, but it is not determined by their extremes, since the part is always less than the whole, *viz.*, the diesis than the greater interval 3 : 4 which exists between 2 : 3 and 1 : 2 and which whole would be here assumed to be determined by the extreme movements.

XXIV. Proposition. *The two planets which shift the kind* (genus) *of harmony, which is the difference between the private ratios of the extreme movements, ought to make a diesis, and the private ratio of one ought to be greater than a diesis, and they ought to make one of the sixths with their aphelial movements and the other with their perihelial.*

For, since the extremes of the movements make two consonances differing by a single diesis, that can take place in three ways. For either the movement of one planet will remain constant and the movement of the other will vary by a diesis, or both will vary by half a diesis and make 3 : 5, a major sixth, when the upper is at its aphelion and the lower in its perihelion, and when they move out of those intervals and advance towards one another, the upper into its perihelion and the lower into its aphelion, they make 5 : 8, a minor sixth; or, finally, one varies its movement from aphelion to perihelion more than the other does, and there is an excess of one diesis, and thus there is a major sixth between the two aphelia, and a minor sixth between the two perihelia. But the first way is not legitimate, for one of these planets would be without eccentricity, contrary to Axiom IV. The second way was less beautiful and less expedient; less beautiful, because less harmonic, for the private ratios of the movements of the two planets would have been out of tune (*inconcinnae*), for whatever is less than a diesis is out of tune; moreover it occasions one single planet to labour under this ill-concordant small difference—except that indeed it could not take place, because in this way the extreme movements would have wandered from the pitches of the system or the notes (*clavibus*) of the musical scale, contrary to Proposition XXII. Moreover, it would have been less expedient, because the sixths would have occurred only at those moments in which the planets would have been at the contrary apsides; there would have been no latitude within which these sixths and the universal consonances related to them could have occurred; accordingly, these universal consonances would have been very rare, with all the (*harmonic*) positions of the planets reduced to the narrow limits of definite and single points on their orbits, contrary to Axiom XIX. Accordingly, the third way remains: that both of the planets should vary their own private movements, but one more than the other, by one full diesis at the least.

XXV. Proposition. *The higher of the planets which shift the kind of harmony ought to have the ratio of its private movements less than a minor whole tone 9 : 10; while the lower, less than a semitone 15 : 16.*

For they will make 3 : 5 either with their aphelial movements or with their perihelial, by the foregoing proposition. Not with their perihelial, for then the ratio of their aphelial movements would be 5 : 8. Accordingly, the lower planet would have its private ratio one diesis more than the upper would, by the same foregoing proposition. But that is contrary to Axiom X. Accordingly, they make 3 : 5 with their aphelial movements, and with their perihelial 5 : 8, which is 24 : 25 less than the other. But if the aphelial movements make 3 : 5, a major sixth, therefore, the aphelial movement of the upper together with the perihelial of the lower will make more than a major sixth; for the lower planet will compound directly its full private ratio.

In the same way, if the perihelial movements make 5 : 8, a minor sixth, the perihelial movement of the upper and the aphelial movement of the lower will make less than a minor sixth; for the lower planet will compound inversely its full private ratio. But if the private ratio of the lower equalled the semitone 15 : 16, then too a perfect fifth could occur over and above the sixths, because the minor sixth, diminished by a semitone, because the perfect fifth; but this is contrary to Proposition XXIII. Accordingly, the lower planet has less than a semitone in its own interval. And because the private ratio of the upper is one diesis greater than the private ratio of the lower, but the diesis compounded with the semitone makes 9 : 10 the minor whole tone.

XXVI. Proposition. *On the planets which shift the kind of harmony, the upper was due to have either a diesis squared, 576 : 625, i.e., approximately 12 : 13, as* [308] *the interval made by its extreme movements, or the semitone 15 : 16, or something intermediate differing by the comma 80 : 81 either from the former or the latter; while the lower planet, either the simple diesis 24 : 25, or the difference between a semitone and a diesis, which is 125 : 128, i.e., approximately 42 : 43; or, finally and similarly, something intermediate differing either from the former or from the latter by the comma 80 : 81, viz., the upper planet ought to make the diesis squared diminished by a comma, and the lower, the simple diesis diminished by a comma.*

For, by Proposition XXV, the private ratio of the upper ought to be greater than a diesis, but by the preceding proposition less than the (minor) whole tone 9 : 10. But indeed the upper planet ought to exceed the lower by one diesis, by Proposition XXIV. And harmonic beauty persuades us that, even if the private ratios of these planets cannot be harmonic, on account of their smallness, they should at least be from among the concordant (*ex concinnis*) if that is possible, by Axiom I. But there are only two concords less than 9 : 10, the (minor) whole tone, *viz.*, the semitone and the diesis; but they differ from one another not by the diesis but by some smaller interval, 125 : 128. Accordingly, the upper cannot have the semitone; nor the lower, the diesis; but either the upper will have the semitone 15 : 16, and the lower, 125 : 128, *i.e.*, 42 : 43; or else the lower will have the diesis 24 : 25, but the upper the diesis squared, approximately 12 : 13. But since the laws of both planets are equal, therefore, if the nature of the concordant had to be violated in their private ratios, it had to be violated equally in both, so that the difference between their private intervals could remain an exact diesis, which is necessary for distinguishing the kinds of consonances, by Proposition XXIV. But the nature of the concordant was then violated equally in both, if the interval whereby the private ratio of the upper planet fell short of the diesis squared and exceeded the semitone is the same interval whereby the private ratio of the lower planet fell short of a simple diesis and exceeded the interval 125 : 128.

Furthermore, this excess or defect was due to be the comma 80 : 81, because, once more, no other interval was designated by the harmonic ratios, and in order that the comma might be expressed among the celestial movements as it is expressed in harmonics, namely, by the mere excess and defect of the intervals in respect to one another. For in harmonies the comma distinguishes between major and minor whole tones and does not appear in any other way.

It remains for us to inquire which ones of the intervals set forth are preferable—whether the diesis, the simple diesis for the lower planet and the diesis squared for the upper, or the semitone for the upper and 125 : 128 for the lower. And the dieses win by the following arguments: For although the semitone has been variously expressed in the musical scale, yet its allied ratio 125 : 128 has not been expressed. On the other hand, the diesis has been expressed variously and the diesis squared somehow, *viz.*, in the resolution of whole tones into dieses, semitones, and lemmas; for then, as has been said in Book III, Chapter 8, two dieses proximately succeed one another in two pitches. The other argument is that in the distinction into kinds, the laws of the diesis are proper but not at all those of the semitone. Accordingly, there had to be greater consideration of the diesis than of the semitone. It is inferred from everything that the private ratio of the upper planet ought to be 2916 : 3125 or approximately 14 : 15, and that of the lower, 243 : 250 or approximately 35 : 36.

It is asked whether the Highest Creative Wisdom has been occupied in making these tenuous little reckonings. I answer that it is possible that many reasons are hidden from me, but if the nature of harmony has not allowed weightier reasons—since we are dealing with ratios which descend below the magnitude of all concords—it is not absurd that God has followed even those reasons, wherever they appear tenuous, since He has ordained nothing without cause. It would be far more absurd to assert that God has taken at random these magnitudes below the limits prescribed for them, the minor whole tone; and it is not sufficient to say: He took them of that magnitude because He chose to do so. For in geometrical things, which are subject to free choice, God chose nothing without a geometrical cause of some sort, as is apparent in the edges of leaves, in the scales of fishes, in the skins of beasts and their spots and the order of the spots, and similar things.

XXVII. Proposition. *The ratio of movements of the Earth and Venus ought to have been greater than a major sixth between the aphelial movements; less than a minor sixth between the perihelial movements.*

By Axiom XX it was necessary to distinguish the kinds of consonances. But by Proposition XXIII that could not be done except through the sixths. Accordingly, since by Proposition XV the Earth and Venus, planets next to one another and icosahedral, had received the minor sixth, 5 : 8, it was necessary for the other sixth, 3 : 5, to be assigned to them, but not between the converging or diverging extremes, but between the extremes of the same field, one sixth [309] between the aphelial, and the other between the perihelial, by Proposition XXIV. Furthermore, the consonance 3 : 5 is cognate to the icosahedron, since both are of the pentagonal cut. See Chapter 2.

Behold the reason why exact consonances are found between the aphelial and perihelial movements of these two planets, but not between the converging, as in the case of the upper planets.

XXVIII. Proposition. *The private ratio of movements fitting the Earth was approximately 14 : 15, Venus, approximately 35 : 36.*

For these two planets had to distinguish the kinds of consonances, by the preceding proposition; therefore, by Proposition XXVI, the Earth as the higher was due to receive the interval 2916 : 3125, *i.e.,* approximately 14 : 15, but Venus as the lower the interval 243 : 250, *i.e.,* approximately 35 : 36.

Behold the reason why these two planets have such small eccentricities and, in proportion to them, small intervals or private ratios of the extreme movements, although nevertheless the next higher planet, Mars, and the next lower, Mercury, have marked eccentricities and the greatest of all. And astronomy confirms the truth of this; for in Chapter 4 the Earth clearly had 14 : 15, but Venus 34 : 35, which astronomical certitude can barely discern from 35 : 36 in this planet.

XXIX. Proposition. *The greater consonance of the movements of Mars and the Earth, viz., that of the diverging movements, could not be from among the consonances greater than 5 : 12.*

Above, in Proposition XVII, it was not any one of the lesser ratios; but now it is not any one of the greater ratios either. For the other common or lesser consonance of these two planets is 2 : 3, when the private ratio of Mars, which by Proposition XIV exceeds 18 : 25, makes more than 12 : 25, *i.e.,* 60 : 125. Accordingly, compound the private ratio of the Earth 14 : 15, *i.e.,* 50 : 60, by the preceding proposition. The compound ratio is greater than 56 : 125, which is approximately 4 : 9, *viz.,* slightly greater than an octave and a major whole tone. But the next greater consonance than the octave and whole tone is 5 : 12, the octave and minor third.

Note that I do not say that this ratio is neither greater nor smaller than 5 : 12; but I say that if it is necessary for it to be harmonic, no other consonance will belong to it.

XXX. Proposition. *The private ratio of movements of Mercury was due to be greater than all the other private ratios.*

For by Proposition XVI the private movements of Venus and Mercury compounded together were due to make about 5 : 12. But the private ratio of Venus, taken separately, is only 243 : 250, *i.e.,* 1458 : 1500. But if it is compounded inversely with 5 : 12, *i.e.,* 625 : 1500, Mercury singly is left with 625 : 1458, which is greater than an octave and a major whole tone; although the private ratio of Mars, which is the greatest of all these among the remaining planets, is less than 2 : 3, *i.e.,* the perfect fifth.

And thereby the private ratios of Venus and Mercury, the lowest planets, if compounded together, are approximately equal to the compounded private ratios of the four higher planets, because, as will now be apparent immediately, the compounded private ratios of Saturn and Jupiter exceed 2 : 3; those of Mars fall somewhat short of 2 : 3: all compounded, 4 : 9, i.e., 60 : 135. Compound the Earth's 14 : 15, i.e., 56 : 60, the result will be 56 : 135, which is slightly greater than 5 : 12, which just now was the compound of the private ratios of Venus and Mercury. But this has not been sought for nor taken from any separate and singular archetype of beauty but comes of itself, by the necessity of the causes bound together by the consonances hitherto established.

XXXI. Proposition. *The aphelial movement of the Earth had to harmonize with the aphelial movement of Saturn, through some certain number of octaves.*

For, by Proposition XVIII, it was necessary for there to be universal consonances, wherefore also there had to be a consonance of Saturn with the Earth and Venus. But if one of the extreme movements of Saturn had harmonized with neither of the Earth's and Venus', this would have been less harmonic than if both of its extreme movements had harmonized with these planets, by Axiom I. Therefore both of Saturn's extreme movements had to harmonize, the aphelial with one of these two planets, the perihelial with the other, since nothing would hinder, as was the case with the first planet. Accordingly these consonances will be either identisonant[1] (*identisonae*) or diversisonant (*diversisonae*), i.e., either of continued double proportion or of some other. But both of them cannot be of some other proportion, for between the terms 3 : 5 (which determine the greater consonance between the aphelial movements of the Earth and Venus, by Proposition XXVII) two harmonic means cannot be set up; for the sixth cannot be divided into three intervals (see Book III). Accordingly, Saturn could not, [310] by means of both its movements, make an octave with the harmonic means between 3 and 5; but in order that its movements should harmonize with the 3 of the Earth and the 5 of Venus, it is necessary that one of those terms should harmonize identically, or through a certain number of octaves, with the others, viz., with one of the said planets. But since the identisonant consonances are more excellent, they had to be established between the more excellent extreme movements, viz., between the aphelial, because too they have the position of a principle on account of the altitude of the planets and because the Earth and Venus claim as their private ratio somehow and as a prerogative the consonance 3 : 5, with which as their greater consonance we are now dealing. For although, by Proposition XXII, this consonance belongs to the perihelial movement of Venus and some intermediate movement of the Earth, yet the start is made at the extreme movements and the intermediate movements come after the beginnings.

1. "Identisonant consonances" are such as 3 : 5, 3 : 10, 3 : 20, etc.

Now, since on one side we have the aphelial movement of Saturn at its greatest altitude, on the other side the aphelial movement of the Earth rather than Venus is to be joined with it, because of these two planets which distinguish the kinds of harmony, the Earth, again, has the greater altitude. There is also another nearer cause: the posterior reasons—with which we are now dealing—take away from the prior reasons but only with respect to minima, and in harmonics that is with respect to all intervals less than concords. But by the prior reasons the aphelial movement not of Venus but of the Earth, will approximate the consonance of some number of octaves to be established with the aphelial movement of Saturn. For compound together, first, 4 : 5 the private ratio of Saturn's movements, *i.e.*, from the aphelion to the perihelial of Saturn (Proposition XI), secondly, the 1 : 2 of the converging movements of Saturn and Jupiter, *i.e.*, from the perihelion of Saturn to the aphelion of Jupiter (by Proposition VIII), thirdly, the 1 : 8 of the diverging movements of Jupiter and Mars, *i.e.*, from the aphelion of Jupiter to the perihelion of Mars (by Proposition XIV), fourthly, the 2 : 3 of the converging movements of Mars and the Earth, *i.e.*, from the perihelion of Mars to the aphelion of the Earth (by Proposition XV): you will find between the aphelion of Saturn and the perihelion of the Earth the compound ratio 1 : 30, which falls short of 1 : 32, or five octaves, by only 30 : 32, *i.e.*, 15 : 16 or a semitone. And so, if a semitone, divided into particles smaller than the least concord, is compounded with these four elements there will be a perfect consonance of five octaves between the aphelial movements of Saturn and the Earth, which have been set forth. But in order for the same aphelial movement of Saturn to make some number of octaves with the aphelial movement of Venus, it would have been necessary to snatch approximately a whole perfect fourth from the prior reasons; for if you compound 3 : 5, which exists between the aphelial movements of the Earth and Venus, with the ratio 1 : 30 compounded of the four prior elements, then as it were from the prior reasons, 1 : 50 is found between the aphelial movements of Saturn and Venus: This interval differs from 1 : 32, or five octaves, by 32 : 50, *i.e.*, 16 : 25, which is a perfect fifth and a diesis; and from six octaves, or 1 : 64, it differs by 50 : 64, *i.e.*, 25 : 32, or a perfect fourth minus a diesis. Accordingly, an indentisonant consonance was due to be established, not between the aphelial movements of Venus and Saturn but between those of Venus and the Earth, so that Saturn might keep a diversisonant consonance with Venus.

XXXII. Proposition. *In the universal consonances of planets of the minor scale the exact aphelial movement of Saturn could not harmonize precisely with the other planets.*

For the Earth by its aphelial movement does not concur in the universal consonance of the minor scale, because the aphelial movements of the Earth and Venus make the interval 3 : 5, which is of the major scale (by Proposition XVII). But by its aphelial movement Saturn makes an identisonant consonance with the aphelial movement of the Earth (by Proposition XXXI). Therefore, neither does Saturn concur by its aphelial movement. Nevertheless, in place of the aphelial movement there follows some faster movement of Saturn, very near to the aphelial, and also in the minor scale—as was apparent in Chapter 7.

XXXIII. Proposition. *The major kind of consonances and musical scale is akin to the aphelial movements; the minor to the perihelial.*

For although a major consonance (*dura harmonia*) is set up not only between the aphelial movement of the Earth and the aphelial movement of Venus but also between the lower aphelial movements and the lower movements of Venus as far as its perihelion; and, conversely, there is a minor consonance not merely between the perihelial movement of Venus and the perihelial of the Earth but also between the higher movements of Venus as far as the aphelion and the higher movements of the Earth (by Propositions XX and XXIV). Accordingly, the major scale is designated properly only in the aphelial movements, the minor, only in the perihelial.

XXXIV. Proposition. *The major scale is more akin to the upper of the two planets, the minor, to the lower.*

[311] For, because the major scale is proper to the aphelial movements, the minor, to the perihelial (by the preceding proposition), while the aphelial are slower and graver than the perihelial; accordingly, the major scale is proper to the slower movements, the minor to the faster. But the upper of the two planets is more akin to the slow movements, the lower, to the fast, because slowness of the private movement always follows upon altitude in the world. Therefore, of two planets which adjust themselves to both modes, the upper is more akin to the major mode of the scale, the lower, to the minor. Further, the major scale employs the major intervals 4 : 5 and 3 : 5, and the minor, the minor ones, 5 : 6 and 5 : 8. But, moreover, the upper planet has both a greater sphere and slower, *i.e.*, greater movements and a lengthier circuit; but those things which agree greatly on both sides are rather closely united.

XXXV. Proposition. *Saturn and the Earth embrace the major scale more closely, Jupiter and Venus, the minor.*

For, first, the Earth, as compared with Venus and as designating both scales along with Venus, is the upper. Accordingly, by the preceding proposition, the Earth embraces the major scale chiefly; Venus, the minor. But with its aphelial movement Saturn harmonizes with the Earth's aphelial movement, through an octave (by Proposition XXXI): wherefore too (by Proposition XXXIII) Saturn embraces the major scale. Secondly, by the same proposition, Saturn by means of its aphelial movement nurtures more the major scale and (by Proposition XXXII) spits out the minor scale. Accordingly, it is more closely related to the major scale than to the minor, because the scales are properly designated by the extreme movements.

Now as regards Jupiter, in comparison with Saturn it is lower; therefore as the major scale is due to Saturn, so the minor is due to Jupiter, by the preceding proposition.

XXXVI. Proposition. *The perihelial movement of Jupiter had to concord with the perihelial movement of Venus in one scale but not also in the same consonance; and all the less so, with the perihelial movement of the Earth.*

For, because the minor scale chiefly was due to Jupiter, by the preceding proposition, while the perihelial movements are more akin to the minor scale (by Proposition XXX), accordingly, by its perihelial movement Jupiter had to designate the key of the minor scale, *viz.*, its definite pitch or key-note (*phthongum*). But too the perihelial movements of Venus and the Earth designate the same scale (by Proposition XXVIII); therefore the perihelial movement of Jupiter was to be associated with their perihelial movements in the same tuning, but it could not constitute a consonance with the perihelial movements of Venus. For, because (by Proposition VIII) it had to make about 1 : 3 with the aphelial movement of Saturn, *i.e.*, the note (*clavem*) d of that system, wherein the aphelial movement of Saturn strikes the note G, but the aphelial movement of Venus the note e: accordingly, it approached the note e within an interval of least consonance. For the least consonance is 5 : 6, but the interval between d and e is much smaller, *viz.*, 9 : 10, a whole tone. And although in the perihelial tension (*tensione*) Venus is raised from the d of the aphelial tension yet this elevation is less than a diesis, (by Proposition XXVIII). But the diesis (and hence any smaller interval) if compounded with a minor whole tone does not yet equal 5 : 6 the interval of least consonance. Accordingly, the perihelial movement of Jupiter could not observe 1 : 3 or thereabouts with the aphelial movement of Saturn and at the same time harmonize with Venus. Nor with the Earth. For if the perihelial movement of Jupiter had been adjusted to the key of the perihelial movement of Venus in the same tension in such fashion that below the quantity of least concord it should preserve with the aphelial movement of Saturn the interval 1 : 3, *viz.*, by differing from the perihelial movement of Venus by a minor whole tone, 9 : 10 or 36 : 40 (besides some octaves) towards the low. Now the perihelial movement of the Earth differs from the same perihelial movement of Venus by 5 : 8, *i.e.*, by 25 : 40. And so the perihelial movements of the Earth and Jupiter differ by 25 : 36, over and above some number of octaves. But that is not harmonic, because it is the square of 5 : 6, or a perfect fifth diminished by one diesis.

XXXVII. Proposition. *It was necessary for an interval equal to the interval of Venus to accede to the 2 : 3 of the compounded private consonances of Saturn and Jupiter and to 1 : 3 the great consonance common to them.*

For with its aphelial movement Venus assists in the proper designation of the major scale; with its perihelial, that of the minor scale, by Propositions XXVII and

XXXIII. But by its aphelial movement Saturn had to be in concord also with the major scale and thus with the aphelial movement of Venus, by Proposition XXXV, but Jupiter's perihelial with the perihelial of Venus, by the preceding proposition. Accordingly, as great as Venus makes its interval from aphelial to perihelial to be, so great an interval must also accede to that movement of Jupiter which makes 1 : 3 with the aphelial movement of Saturn—to the very perihelial movement of Jupiter. But the consonance of the converging movements of Jupiter and Saturn is precisely 1 : 2, by Proposition VIII. Accordingly, if the interval 1 : 2 is divided into the interval [312] greater than 1 : 3, there results, as the compound of the private ratios of both, something which is proportionately greater than 2 : 3.

Above, in Proposition XXVI, the private ratio of the movements of Venus was 243 : 250 or approximately 35 : 36; but in Chapter 4, between the aphelial movement of Saturn and the perihelial movement of Jupiter there was found a slightly greater excess beyond 1 : 3, viz., between 26 : 27 and 27 : 28. But the quantity here prescribed is absolutely equalled, by the addition of a single second to the aphelial movement of Saturn, and I do not know whether astronomy can discern that difference.

XXXVIII. Proposition. *The increment 243 : 250 to 2 : 3, the compound of the private ratios of Saturn and Jupiter, which was up to now being established by the prior reasons, was to be distributed among the planets in such fashion that of it the comma 80 : 81 should accede to Saturn and the remainder, 19,683 : 20,000 or approximately 62 : 63, to Jupiter.*

It follows from Axiom XIX that this was to have been distributed between both planets so that each could with some latitude concur in the universal consonances of the scale akin to itself. But the interval 243 : 250 is smaller than all concords: accordingly no harmonic rules remain whereby it may be divided into two concordant parts, with the single exception of those of which there was need in the division of 24 : 25, the diesis, above in Proposition XXVI; namely, in order that it may be divided into the comma 80 : 81 (which is a primary one of those intervals which are subordinate to the concordant) and into the remainder 19,683 : 20,000, which is slightly greater than a comma, viz., approximately 62 : 63. But not two but one comma had to be taken away, lest the parts should become too unequal, since the private ratios of Saturn and Jupiter are approximately equal (according to Axiom X extended even to concords and parts smaller than those) and also because the comma is determined by the intervals of the major whole tone and minor whole tone, not so two commas. Furthermore, to Saturn the higher and mightier planet was due not that part which was greater, although Saturn had the greater private consonance 4 : 5, but that one which is prior and more beautiful, *i.e.*, more harmonic. For in Axiom X the consideration of priority and harmonic perfection comes first, and the consideration of quantity comes last, because there is no beauty in quantity of itself. Thus the movements of Saturn become 64 : 81, an adulterine[1] major third, as we have called them in Book III, Chapter 12, but those of Jupiter, 6,561 : 8,000.

I do not know whether it should be numbered among the causes of the addition of a comma to Saturn that the extreme intervals of Saturn can constitute the ratio 8 : 9, the major whole tone, or whether that resulted without further ado from the preceding causes of the movements. Accordingly, you here have, in place of a corollary, the reason why, above in Chapter 4, the intervals of Saturn were found to embrace approximately a major whole tone.

XXXIX. Proposition. *Saturn could not harmonize with its exact perihelial movement in the universal consonances of the planets of the major scale, nor Jupiter with its exact aphelial movement.*

For since the aphelial movement of Saturn had to harmonize exactly with the aphelial movements of the Earth and Venus (by Proposition XXXI), that movement of Saturn which is 4 : 5 or one major third faster than its aphelial will also harmonize with them. For the aphelial movements of the Earth and Venus make a major sixth, which, by the demonstrations of Book III, is divisible into a perfect fourth and a major third, therefore the movement of Saturn, which is still faster than this movement already harmonized but none the less below the magnitude of a concordant interval, will not exactly harmonize. But such a movement is Saturn's perihelial movement itself, because it differs from its aphelial movement by more than the interval 4 : 5, *viz.*, one comma or 80 : 81 more (which is less than the least concord), by Proposition XXXVIII. Accordingly the perihelial movement of Saturn does not exactly harmonize. But neither does the aphelial movement of Jupiter do so precisely. For while it does not harmonize precisely with the perihelial movement of Saturn, it harmonizes at a distance of a perfect octave (by Proposition VIII), wherefore, according to what has been said in Book III, it cannot precisely harmonize.

XL. Proposition. *It was necessary to add the lemma of Plato to 1 : 8, or the triple octave, the joint consonance of the diverging movements of Jupiter and Mars established by the prior reasons.*

For because, by Proposition XXXI, there had to be 1 : 32, *i.e.*, 12 : 384, between the aphelial movements of Saturn and the Earth, but there had to be 3:2, *i.e.*, 384 : 256, from the aphelion of the Earth to the perihelion of Mars [313] (by Proposition XV), and from the aphelion of Saturn to its perihelion, 4 : 5 or 12 : 15 with its increment (by Proposition XXXVII); finally, from the perihelion of Saturn to the aphelion of Jupiter 1 : 2 or 15 : 30 (by Proposition VIII); accordingly, there remains

---

1. See footnote to *Intervals Compared with Harmonic Ratios.*

30 : 256 from the aphelion of Jupiter to the perihelion of Mars, by the subtraction of the increment of Saturn. But 30 : 256 exceeds 32 : 256 by the interval 30 : 32, *i.e.*, 15 : 16 or 240 : 256, which is a semitone. Accordingly, if the increment of Saturn, which (by Proposition XXXVIII) had to be 80 : 81, *i.e.*, 240 : 243, is compounded inversely with 240 : 243, the result is 243 : 256; but that is the lemma of Plato,[1] *viz.*, approximately 19 : 20, see Book III. Accordingly, Plato's lemma had to be compounded with the 1:8.

And so the great ratio of Jupiter and Mars, *viz.*, of the diverging movements, ought to be 243 : 2,048, which is somehow a mean between 243 : 2,187 and 243 : 1,944, *i.e.*, between 1 : 9 and 1 : 8, whereof proportionality required the first, above; and a nearer harmonic concord, the second.

XLI. Proposition. *The private ratio of the movements of Mars has necessarily been made the square of the harmonic ratio 5 : 6, viz., 25 : 86.*

For, because the ratio of the diverging movements of Jupiter and Mars had to be 243 : 2,048, *i.e.*, 729 : 6,144, by the preceding proposition, but that of the converging movements 5 : 24, *i.e.*, 1,280 : 6,144 (by Proposition XIII), therefore the compound of the private ratios of both was necessarily 729 : 1,280 or 72,900 : 128,000. But the private ratio of Jupiter alone had to be 6,561 : 8,000, *i.e.*, 104,976 : 128,000 (by Proposition XXVIII). Therefore, if the compound ratio of both is divided by this, the private ratio of Mars will be left as 72,900 : 104,976, *i.e.*, 25 : 36, the square root of which is 5 : 6.

In another fashion, as follows: There is 1 : 32 or 120 : 3,840 from the aphelial movement of Saturn to the aphelial movement of the Earth, but from that same movement to the perihelial of Jupiter there is 1 : 3 or 120 : 360, with its increment. But from this to the aphelial movement of Mars is 5 : 24 or 360 : 1,728. Accordingly, from the aphelial movement of Mars to the aphelial movement of the Earth, there remains 1,728 : 3,840 minus the increment of the ratio of the diverging movements of Saturn and Jupiter. But from the same aphelial movement of the Earth to the perihelial of Mars there is 3 : 2, *i.e.*, 3,840 : 2,500. Therefore between the aphelial and perihelial movements of Mars there remains the ratio 1,728 : 2,560, *i.e.*, 27 : 40 or 81 : 120, minus the said increment. But 81 : 120 is a comma less than 80 : 120 or 2 : 3. Therefore, if a comma is taken away from 2 : 3, and the said increment (which by Proposition XXXVIII is equal to the private ratio of Venus) is taken away too, the private ratio of Mars is left. But the private ratio of Venus is the diesis diminished by a comma, by Proposition XXVI. But the comma and the diesis diminished by a comma make a full diesis or 24 : 25. Therefore if you divide 2 : 3, *i.e.*, 24 : 36 by the diesis 24 : 25, Mars' private ratio of 25 : 36 is left, as before, the square root of which, or 5 : 6, goes to the intervals, by Chapter 3.

1. *Timaeus*, 36.

Behold again the reason why—above, in Chapter 4—the extreme intervals of Mars have been found to embrace the harmonic ratio 5 : 6.

XLII. Proposition. *The great ratio of Mars and the Earth, or the common ratio of the diverging movements, has been necessarily made to be 54 : 125, smaller than the consonance 5 : 12 established by the prior reasons.*

For the private ratio of Mars had to be a perfect fifth, from which a diesis has been taken away, by the preceding proposition. But the common or minor ratio of the converging movements of Mars and the Earth had to be a perfect fifth or 2 : 3, by Proposition XV. Finally, the private ratio of the Earth is the diesis squared, from which a comma is taken away, by Propositions XXVI and XXVIII. But out of these elements is compounded the major ratio or that of the diverging movements of Mars and the Earth—and it is two perfect fifths (or 4 : 9, *i.e.*, 108 : 243) plus a diesis diminished by a comma, *i.e.*, plus 243 : 250; namely, it is 108 : 250 or 54 : 125, *i.e.*, 608 : 1,500. But this is smaller than 625 : 1,500, *i.e.*, than 5 : 12, in the ratio 602 : 625, which is approximately 36 : 37, smaller than 625 : 1,5000, *i.e.*, than 5 : 12, in the ratio 602 : 625, which is approximately 36 : 37, smaller than the least concord.

XLIII. Proposition. *The aphelial movement of Mars could not harmonize in some universal consonance; nevertheless it was necessary for it to be in concord to some extent in the scale of the minor mode.*

For, because the perihelial movement of Jupiter has the pitch *d* of acute tuning in the minor mode, and the consonance 5 : 24 ought to have existed between that and the aphelial movement of Mars, therefore, the aphelial movement of Mars occupies the adulterine pitch of the same acute tuning. I say *adulterine* for, although in Book III, Chapter 12, the adulterine consonances were reviewed and deduced from the composition of systems, certain ones which exist in the simple natural system were omitted. [314] And so, after the line which ends 81 : 120, the reader may add: if you divide into it 4 : 5 or 32 : 40, there remains 27 : 32, the subminor sixth,[1] which exists between *d* and *f* or *c* and *e*[2] or *a* and *c* of even the simple octave. And in the ensuing table, the following should be in the first line; for 5 : 6 there is 27 : 32, which is deficient.

From that it is clear that in the natural system the true note (*clavem*) *f*, as regulated by my principles, constitutes a deficient or adulterine minor sixth with the note *d*. Accordingly since between the perihelial movement of Jupiter set up in the true note *d* and the aphelial movement of Mars there is a perfect minor sixth over and above the double octave, but not the diminished (by Proposition XIII), it follows that with its aphelial movement Mars designates the pitch which is one comma higher than the true note *f*; and so it will concord not absolutely but merely to a certain extent in this scale.

1. Here "sixth" (*sexta*) should probably be "third" (*tertia*). E. C., Jr.
2. *C* and *e* do not produce a subminor third in the "natural system." E. C., Jr.

But it does not enter into either the pure or the adulterine universal harmony. For the perihelial movement of Venus occupies the pitch of *e* in this tuning (*tensionem*). But there is dissonance between *e* and *f*, on account of their nearness. Therefore, Mars is in discord with the perihelial movement of one of the planets, *viz.*, Venus. But too it is in discord with the other movements of Venus; they are diminished by a comma less than a diesis: wherefore, since there is a semitone and a comma between the perihelial movement of Venus and the aphelial movement of Mercury, accordingly, between the aphelion of Venus and the aphelion of Mars there will be a semitone and a diesis (neglecting the octaves), *i.e.*, a minor whole tone, which is still a dissonant interval. Now the aphelial movement of Mars concords to that extent in the scale of the minor mode, but not in that of the major. For since the aphelial movement of Venus concords with the *e* of the major mode, while the aphelial movement of Mars (neglecting the octaves) has been made a minor whole tone higher than *e*, then necessarily the aphelial movement of Mars in this tuning would fall midway between *f* and *f* sharp and would make with *g* (which in this tuning would be occupied by the aphelial movement of the Earth) the plainly discordant interval 25 : 27, *viz.*, a major whole tone diminished by a diesis.

In the same way, it will be proved that the aphelial movement of Mars is also in discord with the movements of the Earth. For because it makes a semitone and comma with the perihelial movement of Venus, *i.e.*, 14 : 15 (by what has been said), but the perihelial movements of the Earth and Venus make a minor sixth 5 : 8 or 15 : 24 (by Proposition XXVII). Accordingly, the aphelial movement of Mars together with the perihelial movement of the Earth (the octaves added to it) will make 14 : 24 or 7 : 12, a discordant interval and one not harmonic, like 7 : 6. For any interval between 5 : 6 and 8 : 9 is dissonant and discordant, as 6 : 7 in this case. But no other movement of the Earth can harmonize with the aphelial movement of Mars. For it was said above that it makes the discordant interval 25 : 27 with the Earth (neglecting the octaves); but all from 6 : 7 or 24 : 28 to 25 : 27 are smaller than the least harmonic interval.

XLIV. Corollary. *Accordingly it is clear from the above Proposition XLIII concerning Jupiter and Mars, and from Proposition XXXIX concerning Saturn and Jupiter, and from Proposition XXXVI concerning Jupiter and the Earth, and from Proposition XXXII concerning Saturn, why—in Chapter 5, above—it was found that all the extreme movements of the planets had not been adjusted perfectly to one natural system or musical scale, and that all those which had been adjusted to a system of the same tuning did not distinguish the pitches (loca) of that system in a natural way or effect a purely natural succession of concordant intervals. For the reasons are prior whereby the single planets came into possession of their single consonances; those whereby all the planets, of the universal consonances; and finally, those whereby the universal consonances of the two modes, the major and the minor: when all those have been posited, an omniform adjustment to one natural system is prevented. But if those causes had not necessarily come first, there is no doubt that either one system and one tuning of it would have embraced the extreme movements of all the planets; or, if there was need of two systems for the two modes of song, the major and minor, the very order of the natural scale would have been expressed not merely in one mode, the major, but also in the remaining minor mode. Accordingly, here in Chapter 5, you have the promised causes of the discords through least intervals and intervals smaller than all concords.*

XLV. Proposition. *It was necessary for an interval equal to the interval of Venus to be added to the common major consonance of Venus and Mercury, the double octave, and also the private consonance of Mercury, which were established above in Propositions XII and XIII by the prior reasons,* [315] *in order that the private ratio of Mercury should be a perfect 5 : 12 and that thus Mercury should with both its movements harmonize with the single perihelial movement of Venus.*

For, because the aphelial movement of Saturn, the highest and outmost planet, circumscribed around its regular solid, had to harmonize with the aphelial movement of the Earth, the highest movement of the Earth, which divides the classes of figures; it follows by the laws of opposites that the perihelial movement of Mercury as the innermost planet, inscribed in its figure, the lowest and nearest to the sun, should harmonize with the perihelial movement of the Earth, with the lowest movement of the Earth, the common boundary: the former in order to designate the major mode of consonances, the latter the minor mode, by Propositions XXXIII and XXXIV. But the perihelial movement of Venus had to harmonize with the perihelial movement of the Earth in the consonance 5 : 3, by Proposition XXVII; therefore too the perihelial movement of Mercury had to be tempered with the perihelial of Venus in one scale. But by Proposition XII the consonance of the diverging movements of Venus and Mercury was determined by the prior reasons to be 1 : 4; therefore, now by these posterior reasons it was to be adjusted by the accession of the total interval of Venus. Accordingly, not from further on, from the aphelion, but from the perihelion of Venus to the perihelion of Mercury there is a perfect double octave. But the consonance 3 : 5 of the converging movements is perfect, by Proposition XV. Accordingly if 1 : 4 is divided by 3 : 5, there remains to Mercury singly the private ratio 5 : 12, perfect too, but not further (by Proposition XVI, through the prior reasons) diminished by the private ratio of Venus.

Another reason. Just as only Saturn and Jupiter are touched nowhere on the outside by the dodecahedron and icosahedron wedded together, so only Mercury is untouched on the inside by these same solids, since they touch Mars on the inside, the Earth on both sides, and Venus on the outside. Accordingly, just as something equal to the private ratio of Venus has been added distributively to the private ratios of movements of Saturn and Jupiter, which are supported by the cube and tetrahedron; so now something as great was due to accede to the private ratio of solitary Mercury, which is comprehended by the associated figures of the cube and tetrahedron; because, as the octahedron, a single figure among the secondary figures, does the job of two among the primary, the cube and tetrahedron (concerning which see Chapter 1), so too among the lower planets there is one Mercury in place of two of the upper planets, *viz.*, Saturn and Jupiter.

Thirdly, just as the aphelial movement of the highest planet Saturn had to harmonize, in some number of octaves, *i.e.*, in the continued double ratio, 1 : 32, with the aphelial movement of the higher and nearer of the two planets which shift the mode of consonance (by Proposition XXXI); so, *vice versa*, the perihelial movement of the lowest planet Mercury, again through some number of octaves, *i.e.*, in the continued double ratio, 1 : 4, had to harmonize with the perihelial movement of the lower and similarly nearer of the two planets which shift the mode of consonance.

Fourthly, of the three upper planets, Saturn, Jupiter, and Mars, the single but extreme movements concord with the universal consonances; accordingly both extreme movements of the single lower planet, *viz.*, Mercury, had to concord with the same; for the middle planets, the Earth and Venus, had to shift the mode of consonances, by Propositions XXXIII and XXXIV.

Finally, in the three pairs of the upper planets perfect consonances have been found between the converging movements, but adjusted (*fermentatae*) consonances between the diverging movements and private ratios of the single planets; accordingly, in the two pairs of the lower planets, conversely, perfect consonances had to be found not between the converging movements chiefly, nor between the diverging, but between the movements of the same field. And because two perfect consonances were due to the Earth and Venus, therefore two perfect consonances were due to Venus and Mercury also. And the Earth and Venus had to receive as perfect a consonance between their aphelial movements as between their perihelial, because they had to shift the mode of their consonance; but Venus and Mercury, as not shifting the mode of their consonance, did not also require perfect consonances between both pairs, the aphelial movements and the perihelial; but there came in place of the perfect consonance of the aphelial movements, as being already adjusted the perfect consonance of the converging movements, so that just as Venus, the higher of the lower planets, has the least private ratio of all the private ratios of movements (by Proposition XXVI), and Mercury, the lower of the lower, has received the greatest ratio of all the private ratios of movements (by Proposition XXX), so too the private ratio of Venus should be the most imperfect of all the private ratios or the farthest removed from consonances, while the private ratio of Mercury should be most perfect of all the private ratios, *i.e.*, an absolute consonance without adjustment, and that finally the relations should be everywhere opposite.

For He Who is before the ages and on into the ages thus adorned the great things of His wisdom: nothing excessive, nothing defective, no room for any censure. How lovely are his works! All things, in twos, one [316] *against one, none lacking its opposite. He has strengthened the goods—adornment and propriety—of each and every one and established them in the best reasons, and who will be satiated seeing their glory?*

XLVI. Axiom. *If the interspacing of the solid figures between the planetary spheres is free and unhindered by the necessities of antecedent causes, then it ought to follow to perfection the proportionality of geometrical inscriptions and circumscriptions, and thereby the conditions of the ratio of the inscribed to the circumscribed spheres.*

For nothing is more reasonable than that physical inscription should exactly represent the geometrical, as the work, its pattern.

XLVII. Proposition. *If the inscription of the regular solids among the planets was free, the tetrahedron was due to touch with its angles precisely the perihelial sphere of Jupiter above it, and with centres of its planes precisely the aphelial sphere of Mars below it. But the cube and the octahedron, each placing its angles in the perihelial sphere of the planet above, were due to penetrate the sphere of the inside planet with the centres of their planes, in such fashion that those centres should turn within the aphelial and perihelial spheres: on the other hand, the dodecahedron and icosahedron, grazing with their angles the perihelial spheres of their planets on the outside, were due not quite to touch with the centres of their planes the aphelial spheres of their inner planets. Finally, the dodecahedral echinus, placing its angles in the perihelial sphere of Mars, was due to come very close to the aphelial sphere of Venus with the midpoints of its converted sides which interdistinguish two solid rays.*

For the tetrahedron is the middle one of the primary figures, both in genesis and in situation in the world; accordingly, it was due to remove equally both regions, that of Jupiter and that of Mars. And because the cube was above it and outside it, and the dodecahedron was below it and within it, therefore it was natural that their inscription should strive for the contrariety wherein the tetrahedron held a mean, and that the one of them should make an excessive inscription, and the other a defective, *viz.*, the one should somewhat penetrate the inner sphere, the other not touch it. And because the octahedron is cognate to the cube and has an equal ratio of spheres, but the icosahedron to the dodecahedron, accordingly, whatever the cube has of perfection of inscription, the same was due to the octahedron also, and whatever the dodecahedron, the same to the icosahedron too. And the situation of the octahedron's similar to the situation of the cube, but that of the icosahedron to the situation of the dodecahedron, because as the cube occupies the one limit to the outside, so the octahedron occupies the remaining limit to the inside of the world, but the dodecahedron and icosahedron are midway: accordingly even a similar inscription was proper, in the case of the dodecahedron, one penetrating the sphere of the inner planet, in that of the icosahedron, one falling short of it.

But the echinus, which represents the icosahedron with the apexes of its angles and the dodecahedron with the bases, was due to fill, embrace, or dispose both regions, that between Mars and the Earth with the dodecahedron as well as that between the Earth and Venus with the icosahedron. But the preceding axiom makes clear which of the opposites was due to which association. For the tetrahedron, which has a rational inscribed sphere, has been allotted the middle position among the primary figures and is surrounded on both sides by figures of incommensurable spheres, whereof the outer is the cube, the inner the dodecahedron, by Chapter 1 of this book. But this geometrical quality, *viz.*, the rationality of the inscribed sphere, represents in nature the perfect inscription of the planetary sphere. Accordingly, the cube and its allied figure have their inscribed spheres rational only in square, *i.e.*, in power alone; accordingly, they ought to represent a semiperfect inscription, where, even if not the extremity of the planetary sphere, yet at least something on the inside and rightfully a mean between the aphelial and perihelial spheres—if that is possible through other reasons—is touched by the centres of the planes of the figures. On the other hand, the dodecahedron and its allied figure have their inscribed spheres clearly irrational both in the length of the radius and in the square; accordingly, they ought to represent a clearly imperfect inscription and one touching absolutely nothing of the planetary sphere, *i.e.*, falling short and not reaching as far as the aphelial sphere of the planet with the centres of its planes.

Although the echinus is cognate to the dodecahedron and its allied figure, nevertheless it has a property similar to the tetrahedron. For the radius of the sphere inscribed in its inverted sides is indeed incommensurable with the radius of the circumscribed sphere, but it is, however, commensurable with the length of the distance between two neighbouring angles. And so the perfection of the commensurability of rays is approximately as great as in the tetrahedron; but elsewhere the imperfection is as great as in the [317] dodecahedron and its allied figure. Accordingly it is reasonable too that the physical inscription belonging to it should be neither absolutely tetrahedral nor absolutely dodecahedral but of an intermediate kind; in order that (because the tetrahedron was due to touch the extremity of the sphere with its planes, and the dodecahedron, to fall short of it by a definite interval) this wedge-shaped figure with the inverted sides should stand between the icosahedral space and the extremity of the inscribed sphere and should nearly touch this extremity—if nevertheless this figure was to be admitted into association with the remaining five, and if its laws could be allowed, with the laws of the others remaining. Nay, why do I say "could be allowed"? For they could not do without them. For if an inscription, which was loose and did not come into contact fitted the dodecahedron, what else could confine that indefinite looseness within the limits of a fixed magnitude, except this subsidiary figure cognate to the dodecahedron and icosahedron, and which comes almost into contact with its inscribed sphere and does not fall short (if indeed it does fall short) any more than the tetrahedron exceeds and penetrates—with which magnitude we shall deal in the following.

This reason for the association of the echinus with the two cognate figures (*viz.*, in order that the ratio of the spheres of Mars and Venus, which they had left indefinite, should be made determinate) is rendered very probable by the fact that 1,000, the semidiameter of the sphere of the Earth, is found to be practically a mean proportional between the perihelial sphere of Mars and the aphelial sphere of Venus; as if the interval, which the echinus assigns to the cognate figures, has been divided between them as proportionally as possible.

XLVIII. Proposition. *The inscription of the regular solid figures between the planetary spheres was not the work of pure freedom; for with respect to very small magnitudes it was hindered by the consonances established between the extreme movements.*

For, by Axioms I and II, the ratio of the spheres of each figure was not due to be expressed immediately by itself, but by means of it the consonances most akin to the ratios of the spheres were first to be sought and adjusted to the extreme movements.

Then, in order that, by Axioms XVIII and XX, the universal consonances of the two modes could exist, it was necessary for the greater consonances of the single pairs to be readjusted somewhat, by means of the posterior reasons. Accordingly, in order that those things might stand, and be maintained by their own reasons, intervals were required which are somewhat discordant with those which arise from the perfect inscription of figures between the spheres, by the laws of movements unfolded in Chapter 3. In order that it be proved and made manifest how much is taken away from the single planets by the consonances established by their proper reasons; come, let us build up, out of them, the intervals of the planets from the sun, by a new form of calculation not previously tried by anyone.

Now there will be three heads to this inquiry: First, from the two extreme movements of each planet the similar extreme intervals between it and the sun will be investigated, and by means of them the radius of the sphere in those dimensions, of the extreme intervals, which are proper to each planet. Secondly, by means of the same extreme movements, in the same dimensions for all, the mean movements and their ratio will be investigated. Thirdly, by means of the ratio of the mean movements already disclosed, the ratio of the spheres or mean intervals and also one ratio of the extreme intervals, will be investigated; and the ratio of the mean intervals will be compared with the ratios of the figures.

As regards the first: we must repeat, from Chapter 3, Article VI, that the ratio of the extreme movements is the inverse square of the ratio of the corresponding intervals from the sun. Accordingly, since the ratio of the squares is the square of the ratio of its sides, therefore, the numbers, whereby the extreme movements of the single planets are expressed, will be considered as squares and the extraction of their roots will give the extreme intervals, whereof it is easy to take the arithmetic mean as the semidiameter of the sphere and the eccentricity. Accordingly the consonances so far established have prescribed.

[318] Planets Props.	Ratios of movements	The roots either prolonged or of their multiples	Therefore the semidiameter of the sphere	Eccentricity	In dimensions whereof the semidiameter of the sphere is 100,000
Saturn by XXXVIII	64 : 81	80 : 90	85	5	5,882
Jupiter by XXXVIII	6,561 : 8,000	81,000 : 89,444	85,222	4,222	4,954
Mars by XLI	25 : 36	50 : 60	55	5	9,091
Earth by XXVIII	2,916 : 3,125	93,531 : 96,825	95,178	1,647	1,730
Venus by XXVIII	243 : 250	9,859 : 10,000	99,295	705	710
Mercury by XLV	5 : 12	63,250 : 98,000	80,625	17,375	21,551

For the second of the things proposed, we again have need of Chapter 3, Article XII, where it was shown that the number which expresses the movement which is as a mean in the ratio of the extremes is less than their arithmetic mean, also less than the geometric mean by half the difference between the geometric and arithmetic means. And because we are investigating all the mean movements in the same dimensions, therefore let all the ratios hitherto established between different twos and also all the private ratios of the single planets be set out in the measure of the least common divisible. Then let the means be sought: the arithmetic, by taking half the difference between the extreme movements of each planet, the geometric, by the multiplication of one extreme into the other and extracting the square root of the product; then by subtracting half the difference of the means from the geometric mean, let the number of the mean movement be constituted in the private dimensions of each planet, which can easily, by the rule of ratios, be converted into the common dimensions.

[319] Therefore, from the prescribed consonances, the ratio of the mean diurnal movements has been found, viz., the ratio between the numbers of the degrees and minutes of each planet. It is easy to explore how closely that approaches to astronomy.

Harmonic ratios of two	Numbers of the extreme movements	Private ratios of the single planets	Continued means of the single planets		Halves of the differences	Number of the mean movement in dimensions	
			Arithmetic	Geometric		Private	Common
ǀ1	♄ 139,968	64 ⎫					
		⎬	72.50	72.00	.25	71.75	156,917
ǀ1	♄ 177,147	81					
2	♃ 354,294	6,561 ⎫					
		⎬	7,280.5	7,244.9	17.8	7,227.1	390,263
ǀ5	♃ 432,000	8,000					
ǀ24	♂ 2,073,600	25 ⎫					
		⎬	30.50	30.00	.25	29.75	2,467,584
ǀ2	♂ 2,985,984	36					
32ǀ3	♁ 4,478,976	2,916 ⎫					
		⎬	3,020.500	3,018.692	.904	3,017.788	4,635,322
ǀ5ǀ ♁	4,800,000	3,125					
ǀ5ǀ8	♀ 7,464,960	243 ⎫					
		⎬	246.500	246.475	.0125	246.4625	7,571,328
ǀ1ǀ3	♀ 7,680,000	250					
ǀ5	☿ 12,800,000	5 ⎫					
		⎬	8.500	7.746	.377	7.369	18,864,680
ǀ4	☿ 30,720,000	12					

The third head of things proposed requires Chapter 3, Article VIII. For when the ratio of the mean diurnal movements of the single planets has been found, it is possible to find the ratio of the spheres too. For the ratio of the mean movements is the $3/2^{th}$ power of the inverse ratio of the spheres. But, too, the ratio of the cube numbers is the $3/2^{th}$ power of the ratio of the squares of those same square roots, given in the table of Clavius, which he subjoined to his *Practical Geometry*. Wherefore, if the numbers of our mean movements (curtailed, if need be, of an equal number of ciphers) are sought among the cube numbers of that table, they will indicate on the left, under the heading of the squares, the numbers of the ratio of the spheres; then the eccentricities ascribed above to the single planets in the private ratio of the semidiameters of each may easily be converted by the rule of ratios into dimensions common to all, so that, by their addition to the semidiameters of the spheres and subtraction from them, the extreme intervals of the single planets from the sun may be established. Now we shall give to the semidiameter of the terrestrial sphere the round number 100,000, as is the practice in astronomy, and with the following design: because this number or its square or its cube is always made up of mere ciphers; and so too we shall raise the mean movement of the Earth to the number 10,000,000,000 and by the rule of ratios make the number of the mean movement of any planet be to the number of the mean movement of the Earth, as 10,000,000,000 is to the new measurement. And so the business can be carried on with only five cube roots, by comparing those single cube roots with the one number of the Earth.

Numbers of the mean movements				Eccentricities				
In the original dimensions	In the new dimensions found in inverse order among the cubes	Numbers of the ratio of the spheres found among the squares	Semi-diameters as above	in dimensions Private as above	Common	Extreme intervals resulting Aphelion	Perihelion
♄ 156,917	29,539,960	9,556	85	5	562	10,118	8,994
♃ 390,263	11,877,400	5,206	85,222	4,222	258	5,464	4,948
♂ 2,467,584	1,878,483	1,523	55	5	138	1,661	1,384
♁ 4,635,322	1,000,000	1,000	95,178	1,647	17	1,017	983
♀ 7,571,328	612,220	721	99,295	705	5	726	716
☿ 18,864,680	245,714	392	80,625	17,375	85	476	308

Accordingly, it is apparent in the last column what the numbers turn out to be whereby the converging intervals of two planets are expressed. All of them approach very near to those intervals, which I found from Brahe's observations. In Mercury alone is there some small difference. For astronomy is seen to give the following intervals to it: 470, 388, 306, all shorter. It seems that the reason for the dissonance may be referred either to the fewness of the observations or to the magnitude of the eccentricity. (See Chapter 3). But I hurry on to the end of the calculation.

For now it is easy to compare the ratio of the spheres of the figures with the ratio of the converging intervals.

[320] For if the semidiameter of the sphere circumscribed around the figure

which is commonly 100,000	becomes:	Then the semidiameter of the sphere or circle inscribed in: instead of:		becomes:	Although by the consonances the interval is:	
In the cube	8,994	♄ (Saturn)	57,735	5,194	Mean	♃ 5,206
In the tetrahedron	4,948	♃ (Jupiter)	33,333	1,649	Aphelial	♂ 1,661
In the dodecahedron	1,384	♂ (Mars)	79,465	1,100	Aphelial	♁ 1,018
In the icosahedron	983	♁ (Earth)	79,465	781	Aphelial	♀ 726
In the echinus	1,384	♂ (Mars)	52,573	728	Aphelial	♀ 726
In the octahedron	716	♀ (Venus)	57,735	413	Mean	☿ 392

which is commonly 100,000	becomes:	Then the semidiameter of the sphere or circle inscribed in: instead of:	becomes:	Although by the consonances the interval is:
In the square in the octahedron	716 ♀ (Venus)	70,711	506	Aphelial ☿ 476
	or 476 ☿ (Mercury)	70,711	336	Perihelial ☿ 308

That is to say, the planes of the cube extend down slightly below the middle circle of Jupiter; the octahedral planes, not quite to the middle circle of Mercury; the tetrahedral, slightly below the highest circle of Mars; the sides of the echinus, not quite to the highest circle of Venus; but the planes of the dodecahedron fall far short of the aphelial circle of the Earth; the planes of the icosahedron also fall short of the aphelial circle of Venus, and approximately proportionally; finally, the square in the octahedron is quite inept, and not unjustly, for what are plane figures doing among solids? Accordingly, you see that if the planetary intervals are deduced from the harmonic ratios of movements hitherto demonstrated, it is necessary that they turn out as great as these allow, but not as great as the laws of free inscription prescribed in Proposition XLV would require: because this κόσμοσ γεωμέτρικοσ (geometrical adornment) of perfect inscription was not fully in accordance with that other κόσμον ἁρμόνικον ἐν-δεχόμενον (possible harmonic adornment)—to use the words of Galen, taken from the epigraph to this Book V. So much was to be demonstrated by the calculation of numbers, for the elucidation of the prescribed proposition.

I do not hide that if I increase the consonance of the diverging movements of Venus and Mercury by the private ratio of the movements of Venus, and, as a consequence, diminish the private ratio of Mercury by the same, then by this process I produce the following intervals between Mercury and the sun: 469,388,307, which are very precisely represented by astronomy. But, in the first place, I cannot defend that diminishing by harmonic reasons. For the aphelial movement of Mercury will not square with that musical scale, nor in the planets which are opposite in the world is the planetary principle (*ratio*) of opposition of all conditions kept. Finally, the mean diurnal movement of Mercury becomes too great, and thereby the periodic time, which is the most certain fact in all astronomy, is shortened too much. And so I stay within the harmonic polity here employed and confirmed throughout the whole of Chapter 9. But none the less with this example I call you all forth, as many of you as have happened to read this book and are steeped in the mathematical disciplines and the knowledge of highest philosophy: work hard and either pluck up one of the consonances applied everywhere, interchange it with some other, and test whether or not you will come so near to the astronomy posited in Chapter 4, or else try by reasons whether or not you can build with the celestial movements something better and more expedient and destroy in part or in whole the layout applied by me. But let whatever pertains to the glory of Our Lord and Founder be equally permissible to you by way of this book, and up to this very hour I myself have taken the liberty of everywhere changing those things which I was able to discover on earlier days and which were the conceptions of a sluggish care or hurrying ardour.

[321] XLIX. Envoi. *It was good that in the genesis of the intervals the solid figures should yield to the harmonic ratios, and the major consonances of two planets to the universal consonances of all, in so far as this was necessary.*

With good fortune we have arrived at 49, the square of 7; so that this may come as a kind of Sabbath, since the six solid eights of discourse concerning the construction of the heavens has gone before. Moreover, I have rightly made an *envoi* which could be placed first among the axioms: because God also, enjoying the works of His creation, "saw all things which He had made, and behold! they were very good."

There are two branches to the *envoi*: First, there is a demonstration concerning consonances in general, as follows: For where there is choice among different things which are not of equal weight, there the more excellent are to be put first and the more vile are to be detracted from, in so far as that is necessary, as the very word ὁ κόσμος, which signifies *adornment*, seems to argue. But inasmuch as life is more excellent than the body, the form than the material, by so much does harmonic adornment excel the geometrical.

For as life perfects the bodies of animate things, because they have been born for the exercise of life—as follows from the archetype of the world, which is the divine essence—so movement measures the regions assigned to the planets, each that of its own planet: because that region was assigned to the planet in order that it should move. But the five regular solids, by their very name, pertain to the intervals of the regions and to the number of them and the bodies; but the consonances to the movements. Again, as matter is diffuse and indefinite of itself, the form definite, unified, and determinant of the material, so too there are an infinite number of geometric ratios, but few consonances. For although among the geometrical ratios there are definite degrees of determinations, formation, and restriction, and no more than three can exist from the ascription of spheres to the regular solids; but nevertheless an accident common to all the rest follows upon even these geometrical ratios: an infinite possible section of magnitudes is presupposed, which those ratios whose terms are mutually incommensurable somehow involve in actuality too. But the harmonic ratios are all rational, the terms of all are commensurable and are taken from a definite and finite species of plane figures. But infinity of section represents the material, while commensurability or rationality of terms represents the form. Accordingly, as material desires the form, as the rough-hewn stone, of a just magnitude indeed, the form of a human body, so the geometric ratios of figures desire the consonances—not in order to fashion and form those consonances, but because this material squares better with this form, this quantity of stone with this statue, even this ratio of regular solids with this consonance—therefore in order so that

80

they are fashioned and formed more fully, the material by its form, the stone by the chisel into the form of an animate being; but the ratio of the spheres of the figure by its own, *i.e.*, the near and fitting, consonance.

The things which have been said up to now will become clearer from the history of my discoveries. Since I had fallen into this speculation twenty-four years ago, I first inquired whether the single planetary spheres are equal distances apart from one another (for the spheres are apart in Copernicus, and do not touch one another), that is to say, I recognized nothing more beautiful than the ratio of equality. But this ratio is without head or tail: for this material equality furnished no definite number of mobile bodies, no definite magnitude for the intervals. Accordingly, I meditated upon the similarity of the intervals to the spheres, *i.e.*, upon the proportionality. But the same complaint followed. For although to be sure, intervals which were altogether unequal were produced between the spheres, yet they were not unequally equal, as Copernicus wishes, and neither the magnitude of the ratio nor the number of the spheres was given. I passed on to the regular plane figures: [322] intervals were formed from them by the ascription of circles. I came to the five regular solids: here both the number of the bodies and approximately the true magnitude of the intervals was disclosed, in such fashion that I summoned to the perfection of astronomy the discrepancies remaining over and above. Astronomy was perfect these twenty years; and behold! there was still a discrepancy between the intervals and the regular solids, and the reasons for the distribution of unequal eccentricities among the planets were not disclosed. That is to say, in this house the world, I was asking not only why stones of a more elegant form but also what form would fit the stones, in my ignorance that the Sculptor had fashioned them in the very articulate image of an animated body. So, gradually, especially during these last three years, I came to the consonances and abandoned the regular solids in respect to minima, both because the consonances stood on the side of the form which the finishing touch would give, and the regular solids, on that of the material—which in the world is the number of bodies and the rough-hewn amplitude of the intervals—and also because the consonances gave the eccentricities, which the regular solids did not even promise—that is to say, the consonances made the nose, eyes, and remaining limbs a part of the statue, for which the regular solids had prescribed merely the outward magnitude of the rough-hewn mass.

Wherefore, just as neither the bodies of animate beings are made nor blocks of stone are usually made after the pure rule of some geometrical figure, but something is taken away from the outward spherical figure, however elegant it may be (although the just magnitude of the bulk remains), so that the body may be able to get the organs necessary for life, and the stone the image of the animate being; so too as the ratio which the regular solids had been going to prescribe for the planetary spheres is

inferior and looks only towards the body and material, it has to yield to the consonances, in so far as that was necessary in order for the consonances to be able to stand
closely by and adorn the movement of the globes.

The other branch of the *envoi*, which concerns universal consonances, has a proof
closely related to the first. (As a matter of fact, it was in part assumed above, in XVIII,
among the Axioms.) For the finishing touch of perfection, as it were, is due rather to
that which perfects the world more; and conversely that thing which occupies a second
position is to be detracted from, if either is to be detracted from. But the universal harmony of all perfects the world more than the single twin consonances of different
neighbouring twos. For harmony is a certain ratio of unity; accordingly the planets are
more united, if they all are in concord together in one harmony, than if each two concord separately in two consonances. Wherefore, in the conflict of both, either one of
the two single consonances of two planets was due to yield, so that the universal harmonies of all could stand. But the greater consonances, those of the diverging movements, were due to yield rather than the lesser, those of the converging movements. For
if the divergent movements diverge, then they look not towards the planets of the given
pair but towards other neighbouring planets, and if the converging movements con-
verge, then the movements of one planet are converging toward the movement of the
other, conversely: for example, in the pair Jupiter and Mars the aphelial movement of
Jupiter verges toward Saturn, the perihelial of Mars towards the Earth: but the perihelial movement of Jupiter verges toward Mars, the aphelial of Mars toward Jupiter.
Accordingly the consonance of the converging movements is more proper to Jupiter
and Mars; the consonance of the diverging movements is somehow more foreign to
Jupiter and Mars. But the ratio of union which brings together neighbouring planets
by twos and twos is less disturbed if the consonance which is more foreign and more
removed from them should be adjusted than if the private ratio should be, *viz.*, the one
which exists between the more neighbouring movements of neighbouring planets.
None the less this adjustment was not very great. For the proportionality has been
found in which may stand the universal consonances of all the planets may exist (and
these in two distinct modes), and in which (with a certain latitude of tuning merely
equal to a comma) may also be embraced the single consonances of two neighbouring
planets; the consonances of the converging movements in four pairs, perfect, of the
aphelial movements in one pair, of the perihelial movements in two pairs, likewise perfect; the consonances of the diverging movements in four pairs, these, however, within the difference of one diesis (the very small interval by which the human voice [323]
in figured song nearly always errs; the single consonance of Jupiter and Mars, this
between the diesis and the semitone. Accordingly it is apparent that this mutual yield-
ing is everywhere very good.)

Accordingly let this do for our *envoi* concerning the work of God the Creator. It now remains that at last, with my eyes and hands removed from the tablet of demonstrations and lifted up towards the heavens, I should pray, devout and supplicating, to the Father of lights: *O Thou Who dost by the light of nature promote in us the desire for the light of grace, that by its means Thou mayest transport us into the light of glory, I give thanks to Thee, O Lord Creator, Who hast delighted me with Thy makings and in the works of Thy hands have I exulted. Behold! now, I have completed the work of my profession, having employed as much power of mind as Thou didst give to me; to the men who are going to read those demonstrations I have made manifest the glory of Thy works, as much of its infinity as the narrows of my intellect could apprehend. My mind has been given over to philosophizing most correctly: if there is anything unworthy of Thy designs brought forth by me—a worm born and nourished in a wallowing place of sins—breathe into me also that which Thou dost wish men to know, that I may make the correction: If I have been allured into rashness by the wonderful beauty of Thy works, or if I have loved my own glory among men, while I am advancing in the work destined for Thy glory, be gentle and merciful and pardon me; and finally deign graciously to effect that these demonstrations give way to Thy glory and the salvation of souls and nowhere be an obstacle to that.*

10. EPILOGUE CONCERNING THE SUN, BY WAY OF CONJECTURE

From the celestial music to the hearer, from the Muses to Apollo the leader of the Dance, from the six planets revolving and making consonances to the Sun at the centre of all the circuits, immovable in place but rotating into itself. For although the harmony is most absolute between the extreme planetary movements, not with respect to the true speeds through the ether but with respect to the angles which are formed by joining with the centre of the sun the termini of the diurnal arcs of the planetary orbits; while the harmony does not adorn the termini, *i.e.*, the single movements, in so far as they are considered in themselves but only in so far as by being taken together and compared with one another, they become the object of some mind; and although no object is ordained in vain, without the existence of some thing which may be moved by it, while those angles seem to presuppose some action similar to our eyesight or at least to that sense-perception whereby, in Book IV, the sublunary nature perceived the angles of rays formed by the planets on the Earth: still it is not easy for dwellers on the Earth to conjecture what sort of sight is present in the sun, what eyes there are, or what other instinct there is for perceiving those angles even without eyes and for evaluating the harmonies of the movements entering into the antechamber of the mind by whatever doorway, and finally what mind there is in the sun. None the less, however those things may be, this composition of the six primary spheres around the sun, cherishing it with their perpetual revolutions and as it were adoring it (just as, separately, four moons accompany the globe of Jupiter, two Saturn, but a single moon by its circuit encompasses, cherishes, fosters the Earth and us its inhabitants, and ministers to us) and this special business of the harmonies, which is a most clear footprint of the highest providence over solar affairs, now being added to that consideration, [324] wrings from me the following confession: not only does light go out from the sun into the whole world, as from the focus or eye of the world, as life and heat from the heart, as every movement from the King and mover, but conversely also by royal law these returns, so to speak, of every lovely harmony are collected in the sun from every province in the world, nay, the forms of movements by twos flow together and are bound into one harmony by the work of some mind, and are as it were coined money from silver and gold bullion; finally, the curia, palace, and praetorium or throne-room of the whole realm of nature are in the sun, whatsoever chancellors, palatines, prefects the Creator has given to nature: for them, whether created immediately from the beginning or to be transported hither at some time, has He made ready those seats. For even this terrestrial adornment, with respect to its principal part, for quite a long while lacked the contemplators and enjoyers, for whom however it had been appointed; and those seats were empty. Accordingly the reflection struck my mind, what did the ancient Pythagoreans in Aristotle mean, who used to call the centre of the world (which they referred to as the "fire" but understood by that the sun) "the watchtower of Jupiter," Διος φυλακὴν; what, likewise, was the ancient interpreter pondering in his mind when he rendered the verse of the Psalm as: "He has placed His tabernacle in the sun."

But also I have recently fallen upon the hymn of Proclus the Platonic philosopher (of whom there has been much mention in the preceding books), which was composed to the Sun and filled full with venerable mysteries, if you excise that one κλῦθ (hear me) from it; although the ancient interpreter already cited has explained this to some extent, *viz.*, in invoking the sun, he understands Him Who has placed His tabernacle in the sun. For Proclus lived at a time in which it was a crime, for which the rulers of the world and the people itself inflicted all punishments, to profess Jesus of Nazareth, God Our Savior, and to contemn the gods of the pagan poets (under Constantine, Maxentius, and Julian the Apostate). Accordingly Proclus, who from his Platonic philosophy indeed, by the natural light of the mind, had caught a distant glimpse of the Son of God, that true light which lighteth every man coming into this world, and who already knew that divinity must never be sought with a superstitious mob in sensible things, nevertheless preferred to seem to look for God in the sun rather than in Christ a sensible man, in order that at the same time he might both deceive the pagans by

honoring verbally the Titan of the poets and devote himself to his philosophy, by draw-
ing away both the pagans and the Christians from sensible beings, the pagans from the
visible sun, the Christians from the Son of Mary, because, trusting too much to the
natural light of reason, he spit out the mystery of the Incarnation; and finally that at
the same time he might take over from them and adopt into his own philosophy whatever the Christians had which was most divine and especially consonant with Platonic
philosophy.[1] And so the accusation of the teaching of the Gospel concerning Christ is
laid against this hymn of Proclus, in its own matters: let that Titan keep as his private
possessions χρῦσα ἡνία (golden reins) and ταμιεῖυν φαοῦς, μεσσατὶην, αἰθερος
ἕδρην, κοδμοῦ κραδιαῖον ἐριφεγγεᾳ κυκλὸν (a treasury of light, a seat at the mid-
part of the ether, a radiant circle at the heart of the world), which visible aspect
Copernicus too bestows upon him; let him even keep his παλιννοστοὺς διφρείς
(cyclical chariot-drivings), although according to the ancient Pythagoreans he does not
possess them but in their place τὸ κέντρον, Διὸς φυλακήν (the centre, the watchtower of Zeus)—which doctrine, misshapen by the forgetfulness of ages, as by a flood,
was not recognized by their follower Proclus; let him also keep his γενεθλὴν
Βλαστησασαν (offspring born) of himself, and whatever else is of nature; in turn, let
the philosophy of Proclus yield to Christian doctrines, [325] let the sensible sun yield
to the Son of Mary, the Son of God, Whom Proclus addresses under the name of the
Titan ζωαρκεὸς, ὦ ἄνα, πηγῆς αὐτὸς ἔχων κλήδα (O lord, who dost hold the key
of the life-supporting spring), and that πάντα τεῆς ἐπλήσας ἐλερσινοοῖο
προνόιης (thou didst fulfill all things with thy mind-awakening foresight), and that
immense power over the μοιράων (fates), and things which were read of in no philosophy before the promulgation of the Gospel,[2] the demons dreading him as their threat-
ening scourge, the demons lying in ambush for souls, ὄφρα ὑφιτενοὺς λαθοῖντο
πατρὸς περιφέγγεος αὐλῆς (in order that they might escape the notice of the light-
filled hall of the lofty father); and who except the Word of the Father is that εἰκὼν
παγγενεταο, θεοῦ, οὐφάευτος ἀπ' ἀρρήτου γενετῆρος παύσατο στοιχείων
ὁρυμάγδος ἐπ ἀλληλοῖσιν ἰόντων (image of the all-begetting father, upon whose
manifestation from an ineffable mother the sin of the elements changing into one
another ceased), according to the following: *The Earth was unwrought and a chaotic
mass, and darkness was upon the face of the abyss, and God divided the light from the darkness, the waters from the waters, the sea from the dry land; and: all things were made by
the very Word.* Who except Jesus of Nazareth the Son of God, ψυχῶν ἀναγωγεύς (the
shepherd of souls), to whom ἱκεσιὴ πολυδάκρυος (the prayer of a tearful suppliant)

---

1. It was the judgment of the ancients concerning his book *Metroace* that in it he set forth, not without divine rapture, his universal doctrine concerning God; and by the frequent tears of the author apparent in it all suspicion was removed from the hearers. None the less this same man wrote against the Christians eighteen epichiremata, to which John Philoponus opposed himself, reproaching Proclus with ignorance of Greek thought, which none the less he had undertaken to defend. That is to say, Proclus concealed those things which did not make for his own philosophy.
2. Nevertheless in Suidas some similar things are attributed to ancient Orpheus, nearly equal to Moses, as if his pupil; see too the hymns of Orpheus, on which Proclus wrote commentaries.

is to be offered, in order that He cleanse us from sins and wash us of the filth τῆς γενεθλῆς (of generation)—as if Proclus acknowledged the forms of original sin—and guard us from punishment and evil, πρηυνῶν θόον ὄμμα δικῆς (by making mild the quick eye of justice), namely, the wrath of the Father? And the other things we read of, which are as it were taken from the hymn of Zacharias (or, accordingly, was that hymn a part of the *Metroace*?) Αχλυν ἀποσκεδάσας ὀλεσίμβροτον ἰολοχεύτον (dispersing the poisonous, man-destroying mist), *viz.*, in order that He may give to souls living in darkness and the shadows of death the φάος ἁγνον (holy light) and ὄλβον ἀστυφελικτὸν ἀπ εὐσεβίνέρατείης (unshaken happiness from lovely piety); for that is to serve God in holiness and justice all our days.

Accordingly, let us separate out these and similar things and restore them to the doctrine of the Catholic Church to which they belong. But let us see what the principal reason is why there has been mention made of the hymn. For this same sun which ὕψοθεν ἀρμνίης ῥῦμα πλοῦσιον ἐξοτεύει (sluices the rich flow of harmony from on high)—so too Orpheus κόσμου τὸν ἐναρμόνιον δρόμον ἕλκων (making move the harmonious course of the world)—the same, concerning whose stock Phoebus about to rise κιθαρῆ ὑπὸ θέσκελα μελπῶν εὐνάξει μεγὰ κῦμα βαρυφλσισβοῖο γενεθλῆς (sings marvellous things on his lyre and lulls to sleep the heavy-sounding surge of generation) and in whose dance Paean is the partner, πλήσας ἁρμονὶης παναπήμονος εὔρεα κόσμν (striking the wide sweep of innocent harmony)—him, I say, does Proclus at once salute in the first verse of the hymn as πῦρος νοεροῦ βασιλέα (king of intellectual fire). By that commencement, at the same time, he indicates what the Pythagoreans understood by the word of fire (so that it is surprising that the pupil should disagree with the masters in the position of the centre) and at the same time he transfers his whole hymn from the body of the sun and its quality and light, which are sensibles, to the intelligibles, and he has assigned to that πῦρ νοερὸς (intellectual fire) of his—perhaps the artisan fire of the Stoics—to that created God of Plato, that chief or self-ruling mind, a royal throne in the solar body, confounding into one the creature and Him through Whom all things have been created. But we Christians, who have been taught to make better distinctions, know that this eternal and uncreated "Word," Which was "with God" and Which is contained by no abode, although He is within all things, excluded by none, although He is outside of all things, took up into unity of person flesh out of the womb of the most glorious Virgin Mary, and, when the ministry of His flesh was finished, occupied as His royal abode the heavens, wherein by a certain excellence over and above the other parts of the world, *viz.*, through His glory and majesty, His celestial Father too is recognized to dwell, and has also promised to His faithful, mansions in that house of His Father: as for the remainder concerning that abode, we believe it superfluous to inquire into it too curiously or to forbid the senses or natural reasons to investigate that which the eye has not seen nor the ear heard and into which the heart of man has not ascended; but we duly subordinate the created mind—of whatsoever excellence it may be—to Its Creator, and we introduce neither God-intelligences with Aristotle and the pagan philosophers nor armies of innumerable planetary spirits with the Magi, nor do we propose that they are either to be adored or summoned to intercourse with us by theurgic superstitions, for we have a careful fear of that; but we freely inquire by natural reasons what sort of thing each mind is, especially if in the heart of the world [326] there is any mind bound rather closely to the nature of things and performing the function of the soul of the world— or if also some intelligent creatures, of a nature different from human perchance do inhabit or will inhabit the globe thus animated (see my book *on the New Star*, Chapter 24, "On the Soul of the World and Some of Its Functions"). But if it is permissible, using the thread of analogy as a guide, to traverse the labyrinths of the mysteries of nature, not ineptly, I think, will someone have argued as follows: The relation of the six spheres to their common centre, thereby the centre of the whole world, is also the same as that of διανοὶα (discursive intellection) to νοῦς (intuitive intellection), according as those faculties are distinguished by Aristotle, Plato, Proclus, and the rest; and the relation of the single planets' revolutions in place around the sun to the ἀμετᾴθεδον (unvarying) rotation of the sun in the central space of the whole system (concerning which the sun-spots are evidence; this has been demonstrated in the *Commentaries on the Movement of Mars*) is the same as the relation of τὸ διανοητικὸν to τὸ νοερὸν, that of the manifold discourses of ratiocination to the most simple intellection of the mind. For as the sun rotating into itself moves all the planets by means of the form emitted from itself, so too—as the philosophers teach—mind, by understanding itself and in itself all things, stirs up ratiocinations, and by dispersing and unrolling its simplicity into them, makes everything to be understood. And the movements of the planets around the sun at their centre and the discourses of ratiocinations are so interwoven and bound together that, unless the Earth, our domicile, measured out the annual circle, midway between the other spheres—changing from place to place, from station to station—never would human ratiocination have worked its way to the true intervals of the planets and to the other things dependent from them, never would it have constituted astronomy. (See the *Optical Part of Astronomy*, Chapter 9.)

On the other hand, in a beautiful correspondence, simplicity of intellection follows upon the stillness of the sun at the centre of the world, in that hitherto we have always worked under the assumption that those solar harmonies of movements are defined neither by the diversity of regions nor by the amplitude of the expanses of the world. As a matter of fact, if any mind observes from the sun those harmonies, that mind is without the assistance afforded by the movement and diverse stations of his abode, by means of which it may string together ratiocinations and discourse necessary for measuring out the planetary intervals. Accordingly, it compares the diurnal movements of each planet, not as they are in their own orbits but as they pass through the angles at the centre of the sun. And so if it has knowledge of the magnitude of the spheres, this knowledge must be present in it *a priori*, without any toil of ratiocination: but to what extent that is true of human mind and of sublunary nature has been made clear above, from Plato and Proclus.

Under these circumstances, it will not have been surprising if anyone who has been thoroughly warmed by taking a fairly liberal draft from that bowl of Pythagoras which Proclus gives to drink from in the very first verse of the hymn, and who has been made drowsy by the very sweet harmony of the dance of the planets begins to dream (by telling a story he may imitate Plato's Atlantis and, by dreaming, Cicero's Scipio): throughout the remaining globes, which follow after from place to place, there have been disseminated discursive or ratiocinative faculties, whereof that one ought assuredly to be judged the most excellent and absolute which is in the middle position among those globes, *viz.*, in man's earth, while there dwells in the sun simple intellect, πῦρ νοερὸν, or νοῦς, the source, whatsoever it may be, of every harmony.

For if it was Tycho Brahe's opinion concerning that bare wilderness of globes that it does not exist fruitlessly in the world but is filled with inhabitants: with how much greater probability shall we make a conjecture as to God's works and designs even for the other globes, from that variety which we discern in this globe of the Earth. For He Who created the species which should inhabit the waters, beneath which however there is no room for the air [327] which living things draw in; Who sent birds supported on wings into the wilderness of the air; Who gave white bears and white wolves to the snowy regions of the North, and as food for the bears the whale, and for the wolves, birds' eggs; Who gave lions to the deserts of burning Libya and camels to the widespread plains of Syria, and to the lions an endurance of hunger, and to the camels an endurance of thirst: did He use up every art in the globe of the Earth so that He was unable, every goodness so that he did not wish, to adorn the other globes too with their fitting creatures, as either the long or short revolutions, or the nearness or removal of the sun, or the variety of eccentricities or the shine or darkness of the bodies, or the properties of the figures wherewith any region is supported persuaded?

Behold, as the generations of animals in this terrestrial globe have an image of the male in the dodecahedron, of the female in the icosahedron—whereof the dodecahedron rests on the terrestrial sphere from the outside and the icosahedron from the inside: what will we suppose the remaining globes to have, from the remaining figures? For whose good do four moons encircle Jupiter, two Saturn, as does this our moon this our domicile? But in the same way we shall ratiocinate concerning the globe of the sun also, and we shall as it were incorporate conjectures drawn from the harmonies, *et cetera*—which are weighty of themselves—with other conjectures which are more on the side of the bodily, more suited for the apprehension of the vulgar. Is that globe empty and the others full, if everything else is in due correspondence? If as the Earth breathes forth clouds, so the sun black smoke? If as the Earth is moistened and grows under showers, so the sun shines with those combusted spots, while clear flamelets sparkle in its all fiery body. For whose use is all this equipment, if the globe is empty? Indeed, do not the senses themselves cry out that fiery bodies dwell here which are receptive of simple intellects, and that truly the sun is, if not the king, at least the queen πῦρος νοεροῦ (of intellectual fire)?

Purposely I break off the dream and the very vast speculation, merely crying out with the royal Psalmist: *Great is our Lord and great His virtue and of His wisdom there is no number: praise Him, ye heavens, praise Him, ye sun, moon, and planets, use every sense for perceiving, every tongue for declaring your Creator. Praise Him, ye celestial harmonies, praise Him, ye judges of the harmonies uncovered* (and you before all, old happy Mastlin, for you used to animate these cares with words of hope): *and thou my soul, praise the Lord thy Creator, as long as I shall be: for out of Him and through Him and in Him are all things,* καὶ τὰ αἰσθητὰ καὶ τὰ νοερὰ (*both the sensible and the intelligible*); *for both whose whereof we are utterly ignorant and those which we know are the least part of them; because there is still more beyond. To Him be praise, honour, and glory, world without end. Amen.*

THE END

This work was completed on the 17th or 27th day of May, 1618; but Book V was reread (while the type was being set) on the 9th or 19th of February, 1619. At Linz, the capital of Austria—above the Enns.

Acknowledgments

This book would not have been possible without the help of a number of talented people who made different contributions at various stages of the book's development. Among those deserving special thanks are Michael Rosin, a consultant to Running Press, Gil King, and Mrs. Karen Sime, assistant to Professor Stephen Hawking.

Thanks are also due to several past and present members of the staff of Running Press: Carlo DeVito, Kathleen Greczylo, Kelly Pennick, Bill Jones, Deborah Grandinetti, and Sarah O'Brien.

About Stephen Hawking

Stephen Hawking is considered the most brilliant theoretical physicist since Einstein. He has also done much to popularize science. His book, *A Brief History of Time*, sold more than 10 million copies in 40 languages, achieving the kind of success almost unheard of in the history of science writing. His subsequent books, *The Universe in A Nutshell* and *The Future of Spacetime* with Kip S. Thorne and others, have also been well received.

He was born in Oxford, England, on January 8, 1942 (300 years after the death of Galileo). He studied physics at University College, Oxford, received his PhD in Cosmology at Cambridge, and since 1979 has held the post of Lucasian Professor of Mathematics. The chair was founded in 1663 with money left in the will of the Reverend Henry Lucas, who had been the Member of Parliament for the University. It was first held by Isaac Barrow, and then in 1663 by Isaac Newton. It is reserved for those individuals considered the most brilliant thinkers of their time.

Professor Hawking has worked on the basic laws that govern the universe. With Roger Penrose, he showed that Einstein's General Theory of Relativity implied space and time would have a beginning in the Big Bang and an end in black holes. The results indicated it was necessary to unify General Relativity with Quantum Theory, the other great scientific development of the first half of the twentieth century. One consequence of such a unification that he discovered was that black holes should not be completely black but should emit radiation and eventually disappear. Another conjecture is that the universe has no edge or boundary in imaginary time.

Stephen Hawking has twelve honorary degrees and is the recipient of many awards, medals, and prizes. He is a Fellow of the Royal Society and a Member of the United States National Academy of Sciences. He continues to combine family life (he has three children and one grandchild) and his research into theoretical physics together with an extensive program of travel and public lectures.